HIDDEN WARBIRDS II

More Epic Stories of Finding,
Recovering, and Rebuilding
WWII's Lost Aircraft

Nicholas A. Veronico

ZENITH PRESS

For Betty, with love.

For those who served.

First published in 2014 by Zenith Press, an imprint of Quayside Publishing Group Inc., 400 First Avenue North, Suite 400, Minneapolis, MN 55401 USA

© 2014 Zenith press

Text © 2014 Nicholas A. Veronico

Unless noted otherwise, all archival photos are from the author's collection and all contemporary photos were taken by the author.

The information in this book is true and complete to the best of our knowledge. All recommendations are made without any guarantee on the part of the author or Publisher, who also disclaims any liability incurred in connection with the use of this data or specific details.

We recognize, further, that some words, model names, and designations mentioned herein are the property of the trademark holder. We use them for identification purposes only. This is not an official publication.

Zenith Press titles are also available at discounts in bulk quantity for industrial or sales-promotional use. For details write to Special Sales Manager at Quayside Publishing Group Inc., 400 First Avenue North, Suite 400, Minneapolis, MN 55401 USA.

To find out more about our books, visit us online at www.zenithpress.com.

Library of Congress Cataloging-in-Publication Data

Veronico, Nick, 1961-
  Hidden warbirds II : more epic stories of finding, recovering, and rebuilding WWII's lost aircraft / Nicholas A. Veronico.
    pages cm
  ISBN 978-0-7603-4601-3 (hardback)
  1. Airplanes, Military—Conservation and restoration—United States. 2. Airplanes, Military—Accidents—United States—History—20th century. 3. World War, 1939-1945—Aerial operations. I. Title.
  UG1243.V4692 2014
  940.54'4973—dc23
                              2014001453

Editors: Erik Gilg, Madeleine Vasaly
Design Manager: James Kegley
Cover Designer: Jason Gabbert
Layout Designer: Chris Fayers

On the front cover: On June 11, 2013, the Do-17Z was raised from the sea in an area known as Goodwin Sands, approximately four miles from the English coast. Divers attached lifting straps to the aircraft and it was raised from the bottom and lifted onto a barge. *Courtesy RAF Museum*

On the frontis: Pilot Herm Gallaher set Eight-Zulu down on a Tanana River sandbar after losing power while climbing out from Fairbanks, Alaska, with a load of retardant on July 27, 1969. By 1986, the B-25 was hidden from sight and could only be seen from the air, by flying directly over the bomber. *Milo Peltzer*

Printed in China

10 9 8 7 6 5 4 3 2 1

Contents

Acknowledgments

Hidden Warbirds II is a celebration of those who locate, recover, and restore lost World War II warbirds and the men and women who flew them during the war. More than seventy years have passed since the beginning of the first truly global war in which thousands of aircraft were built, flown, and fought in aerial battlefields the world over. When the war ended, many aircraft were abandoned or unaccounted for, and many have been returned and restored.

Documenting the stories of these aircraft and those involved requires the generosity of many people. And to them, I owe a debt of gratitude and extend my thanks to: Ian Abbott; Carlos Abella; Mark R. Allen; Bill Allmon; Peter Arnold; Jim Azelton; Rebecca Azelton; Brian Baker; Gerald Balzer; Caroline and Ray Bingham; Bruce Brockhagen; Don Brooks; Roger Cain; Patrick Carry and Christopher Johnson; James Carter and Karl von Moller—*Broken Wings*; Tom Camp; Wayne Cook; Lex Crawley; Charles Darby; Ed Davies; John Davis; Robert F. Dorr; Bill Dunbar; Doug Kirby; Shelley Ragsdale; Paul Schumacher; Drannon Vines—National Naval Aviation Museum; Jim Dunn; Ann Y. Evans; Chad Ezell; Nelson Ezell; Randy Ferris; Guttorm Fjeldstad; Curtis Fowles; Craig Fuller—Aviation Archaeological Investigation and Research; Chuck Geise; Erik Gilg—and the staff at Zenith Books; Wayne Gomes; Cory Graff; Kevin Grantham; Woody Grantham; Todd Hackbarth; Dan Hagedorn; Alice Hendricks; Hélio Higuchi; Ted Holgerson—California Airframe Parts; Frederick Johnsen; Vallarie Kilkenney-Jukes and Norm Jukes; Larry Kotz; Robert Kropp; Dr. Gary Kuhn; Martin Kyburz; Christer Landberg; Tillie and William T. Larkins; Jim Larsen; Dave Leininger; Gerry Liang; Roger Mecca; Dale Messimer; Robert Mester; Patrick Mihalek; Chris Miller; John Muszala and John Muszala II; Anne Murata and James Koivunen—Pacific Aviation Museum Pearl Harbor; Robert Nishimura; Michael O'Leary; Mike Oliver—Tillamook Air Museum; Allan Olson and Taras Lyssenko—A. and T. Recovery; Milo Peltzer; Jakub Perka; Dick Phillips; Tyler Pinkerton; Chris Prevost and Sheryl Carlucci; Paul Quinn; Taigh Ramey—Vintage Aircraft; Scott Rose, editor of the Warbird Information Exchange (WIX) as well as many WIXers; Lee Scales; Pat Scannon; Doug Scroggins; Joe Shoen; Jim Slattery; Scott Slocum; Robert Smith; Ajay Srivastava and Michelle Morgans—RAF Museum; Bill Stanczak, Ron Strong; Scott Thompson; Dave Trojan; Rick Turner; Anders Utgaard; Mike Vadeboncoeur—Midwest Aero; Richard VanderMeulen; Paul Varga; Armand and Karen Veronico; Betty Veronico; Tony Veronico; Mike Walton; Mark Watt; Dan F. Webb; Tom Wilson; and Christen Wright.

With great appreciation.
Nicholas A. Veronico
San Carlos, California

Introduction
They're Still Out There . . .

Above and over: Flight Sergeant Dennis Copping belly-landed RAF Kittyhawk ET574 on June 28, 1942, two hundred miles from the nearest settlement in Egypt. Copping was listed as missing in action and his aircraft was not found until Jakub Perka stumbled across it seventy years later. The aircraft is a time capsule and is nearly complete. This aircraft has been recovered by the RAF Museum and is now in storage. *Jakub Perka*

Today, nearly seventy years after the end of World War II, there is a strong movement to find, restore, and fly aircraft from that conflict. It is one thing to see a Messerschmitt Bf 109, a North American Aviation P-51 Mustang, or a Supermarine Spitfire sitting in a museum, but it is another sensory experience to hear the roar of a Rolls-Royce Merlin or Daimler-Benz engine as it pulls a fighter through the air at speeds approaching 400 mph. The sound is heard and the engine's vibrations are felt, depending upon where one stands, and the auditory sensations of the aircraft and its engine bring history to life.

7

It is the sights and sounds of warbirds of all types that have made the flying airshow so popular. All of today's warbird airshows owe their existence to the National Air Races, held at Cleveland, Ohio, before and after World War II. In fact, the warbird movement as a whole owes a debt to the National Air Races.

The war in Europe officially ended on May 7, 1945, punctuated by Adolf Hitler's suicide on April 30, 1945. The Japanese surrendered unconditionally to Allied Forces on September 2, 1945, after two atomic bombs were dropped on its cities (Hiroshima on August 6 and Nagasaki on August 9, 1945), ending the conflict. In January 1946, as servicemen and women were still in the process of returning from duty overseas, it was announced that the National Air Races would return to Cleveland, Ohio, in September of that year.

Servicemen and women had flown aircraft much faster than those that had competed in the pre-war National Air Races, and the most competitive of the ex-military pilots made their way to aircraft surplus yards around the county to secure an aircraft capable of winning the races. The women, most former Women Airforce Service Pilots (WASPs), would compete in the Halle Trophy races, sponsored by Halle department stores. The Halle Trophy Race for women was confined to AT-6/SNJ-type aircraft and consisted of five laps of a fifteen-mile course, or seventy-five miles. Men and women could also compete in the Bendix "R" Division, a cross-country speed dash or any of the various closed-course trophy races. For the cross-country and closed-course races, competitors brought a variety of war-surplus P-38 Lightnings, P-51As, Bs, Cs, and Ds, along with a number of Bell P-39 Airacobras and P-63 Kingcobras.

Because of their value, everyone is keeping an eye out for hidden warbirds. This infrared sensor photo shows a Hawker Fury in Iraq in 2011. Note the five-blade propeller and detached outer wing panels. If it could be exported, it would be worth more than $300,000 as is once it reached the United States or United Kingdom. *Tony Veronico*

Seeing the former combatants competing for prize money enthralled the general public, giving them a sense of what these aircraft did in the skies above Europe and the Pacific. When the U.S. Navy fielded an aerial demonstration team of its own, the Blue Angels, flying first Grumman F6F Hellcats and later F8F Bearcats, they became one of the biggest draws at aviation events around the country. Airshows were front-page news and the flyers who put these war-surplus airplanes through their paces were treated like celebrities. Keep in mind that the Wright Brothers had flown less than fifty years prior and Charles "Lucky Lindy" Lindbergh's solo transatlantic flight had occurred less than two decades before.

If the glamorous air races and post-war employment of a number of surplus World War II aircraft types was the first era of the warbird movement, the second era, essentially from 1950 to 1964, was not so kind. Bombers, fighters, and trainers that had managed to escape the scrappers' torch now sat on airfields across the country, most neglected, too expensive to maintain, or because the owners simply lost interest. The Korean War and the jet age had snuffed out the romance of World War II flight and the escapades of its aircrews. As derelict warbirds gathered dirt and bird droppings across the United States, a small cadre of forward-thinking individuals began to buy up, take home, and rescue a number of ex-World War II warbirds. Simultaneously, the last of the World War II types in the American air arsenal's inventory were being sold off—P-51s went at auction at McClellan AFB, California, and Bearcats, Corsair, Hellcats, Tigercats, and many others were sold surplus at the Navy's storage pool at Litchfield Park,

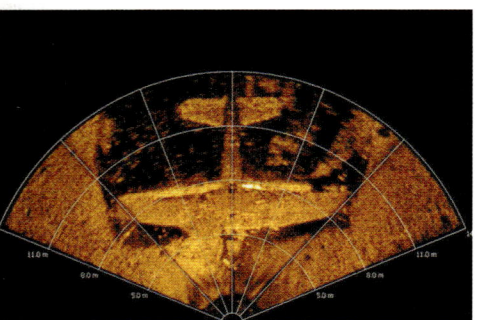

Sonar image of the Grumman F6F Hellcat discovered in 185 feet of water off the coast of Jupiter, Florida. Researchers from OceanGate Inc., were testing a new submersible fitted with 2D and 3D high-frequency sonar when they discovered the sunken fighter. *OceanGate Inc.*

Arizona. Some of the aircraft acquired from these surplus stocks went south of the border only to re-emerge decades later to fuel the warbird movement in the 1970s and 1980s, and others flew aerial mapping missions or attempted to change the weather, flying cloud-seeding missions.

While government aircraft sales were occurring on a small scale, simmering in the background of the 1950s and 1960s was an aviation movement that would eventually explode and bring with it a new enthusiasm for warbirds. That grassroots energy was the homebuilt movement, driven in large part by the Experimental Aircraft Association. Many World War II veterans wanted to fly, but in the 1950s and 1960s were busy becoming educated, building careers, and raising families. They could, however, afford to satisfy their aviation wanderlust by building an aircraft in the garage or in a barn. Later, when some of the group had acquired greater financial means, coupled with the knowledge gained through military service, they began acquiring surplus and former derelict World War II warbirds.

High-performance warbird acquisition by a growing number of aviation enthusiasts, coupled with the fact that World War II was nearly twenty years behind them, saw the third era of the warbird movement come into play. The National Air Races restarted at Reno in 1964, and by the end of the decade, filmmakers were scouring the countryside for pilots and aircraft to fly in such productions as *Catch-22, Tora! Tora! Tora!*, and *The Battle of Britain*.

The fourth era of the warbird movement began in the mid-1970s and lasted until approximately 1992. This was the time of the mass warbird recoveries from Latin and South America, and from a number of islands in the Pacific. P-51s came back to the United States from Australia, Bolivia, the Dominican Republic, Guatemala, and Indonesia, while P-39s and A-20s were hauled out of the jungles of the Pacific, and Republic P-47 Thunderbolts were brought back from Brazil, Hungary, and Peru.

Warbirds entered the fifth era in the early 1990s when the authenticity of the original aircraft became a serious part of the competition.

Using a target list provided by the National Oceanic and Atmospheric Administration and judging by its size, researchers thought they would find a small boat on the bottom. Instead, they found this Hellcat, inverted, showing its distinctive rearward retracting main landing gear. *OceanGate Inc.*

No longer was fogging a fighter's interior with gray paint considered acceptable. The warbird restorers' marching orders were to "put it back to stock"—duplicating stencils correctly, matching original factory paint; and preserving the names, drawings, and other "graffiti" left by war workers on the insides of aircraft. Planes like Butch Schroder's F-6D *Lil' Margaret* are a prime example of the attention to detail that was needed to be recognized as a top-level, authentic restoration during this era.

The sixth era of warbirds, which continues today, has taken warbird restoration up a few notches. The dotcom boom and the

On February 12, 1942, an F2A-3 was ditched into the lagoon at Midway Atoll four months before the June 1942 battle that saw American soldiers, sailors, airmen, and Marines repulse an attempted Japanese invasion. In July 2012, a team from the National Oceanic and Atmospheric Administration discovered the wreckage of F2A-3 Buno 01554 in the atoll's lagoon and documented the wreckage. Pictured is F2A-3 Buno 01516, the first F2A-3 and a sister-ship of the crashed Buno 01554. *NASA Ames Research Center/AILS*

digital revolution had a hand in this. Changes in communication technology alone have done amazing things for the warbird movement. No longer is a researcher sitting at the National Archives pouring over document upon document, requesting copies, or having a photo reproduced, then traveling home to wait a couple of months

until his copies arrive before sharing the research with craftsmen in the shop to "set the standard" for a restoration. Now a researcher can find a document, scan it, and email or text it back to the craftsmen to show exactly how some part or assembly came from the factory, thereby adding to the restoration's authenticity.

The digital revolution brought the Internet, with easy-to-host websites and services like eBay. The Internet has become a way for dedicated individuals to post and share information about warbirds: everything from production drawings to original manuals, to photos on the production line or in combat theaters. The success of many sites' message boards is a testament to how well and quickly knowledge can and is being shared.

A lot of people like to rip into eBay, or "e-Pay" as many call it, but the online marketplace has brought many rare documents out from private collections and the estates of World War II veterans and homefront workers, into the hands of restoration shops and those who specialize in the minutiae of certain aircraft.

The cost of other technologies has also been reduced, for example, the computer controlled numeric milling machine, or CNC machine. Now a warbird part that used to take days or weeks to make can be 3D rendered and uploaded to the CNC machine, and a complex part can be made in a matter of hours.

Composite technologies have also come into their own in the past decade. This has enabled more robust, yet less expensive molds to be made for replicating compound skin structures and other parts. These technologies mean that an aircraft can be taken apart and rebuilt to fly, better than brand new. "Taking an aircraft down to its last rivet" was a cliché at one time; now it's the standard of the warbird movement. Readers will learn about a number of these types of restorations in the coming pages.

Always More to Find

Seventy years after World War II turned in the Allies' favor, one would think that all of the aircraft from those battles had been accounted for. Not so. The years 2012–2014 have been filled with World War II aircraft discoveries and recoveries.

Drawing many parallels to the famous Libyan desert wreck of the B-24 *Lady Be Good*, one of the most intriguing finds of late was Flight Sgt. Dennis Copping's Curtiss Kittyhawk, Royal Air Force (RAF) serial ET574 (ex-U.S. Army Air Forces [USAAF] P-40E 41-35928). This aircraft was received by the USAAF on February 28, 1942, and sent by ship from New York to the British in the Middle East, entering service with 260 Squadron in Egypt.

Four months after delivery, on June 28, 1942, Flight Sgt. Copping was reported missing while ferrying the Curtiss fighter to a repair depot behind the lines in Egypt. Copping and his aircraft were never found during the battle or in any of the numerous after-conflict battlefield sweeps to locate the missing and reinter war dead in national cemeteries. Fast forward to March 2012, when Jakub Perka, a Polish oil company surveyor, located the fighter in the Saharan Desert nearly two hundred miles from the closest town.

To searchers, it appears that Copping successfully belly-landed the Kittyhawk on rough, rock-strewn terrain. He then made a shelter using his parachute and tried to get the radio working. At some point he walked away from the aircraft and his remains have yet to be discovered; however, his aircraft is a virtual time capsule. It was found nearly intact, with the propeller the only major assembly separated from the aircraft.

The RAF Museum moved quickly to recover the aircraft and in summer 2012, contracted with Tim Manna's Kennet Aviation to bring the Kittyhawk home to the United Kingdom. Today, Kittyhawk ET574 is held by the museum in storage awaiting a display space. In exchange for delivering the Kittyhawk, Kennet Aviation received a Spitfire project that was surplus to the museum's collection.

Other desert finds include six Italian Meridionali IMAM Ro.37 reconnaissance/close air support biplanes that were located northeast of Kabul, Afghanistan in 2006, and a Hawker Fury in Iraq. Sixteen Ro.37bis *Lince* (Lynx) were sold to Afghanistan in 1937, and flown until 1941. During a patrol by troops of the Italian Army's 132nd Armored Artillery Regiment, six of the biplanes were discovered in a military junkyard. The aircraft were shipped back to Italy and one

A six-engine Messerschmitt Me-323 was shot down off the coast of Sardina on July 26, 1943, and was recently discovered by sport divers. The aircraft found was shot down only days before the pictured Me-323 was caught skimming the sea near Cape Corso, Sardina. *U.S. Army Air Forces*

of the Ro.37s is nearing the end of its restoration. When complete, it will be displayed at the Italian Air Force Museum at Vigna di Valle, northwest of Rome.

In neighboring Iraq, a Hawker Fury of post–World War II vintage—a Sea Fury without an arresting hook and other gear necessary for aircraft carrier operations—was recently spotted. Egypt, Iraq, and Pakistan acquired a number of the type in the late 1940s, and Iraq used the Fury to continue its suppression of the Kurdish people in the northern part of the country. The Iraqi Air Force aircraft, known as the "Baghdad Furies," were phased out of service by 1960 and replaced with de Havilland Venoms and Hawker Hunters. The majority of the Baghdad Furies were sold to collectors Ed Jurist and David Tallichet in 1978–1979, who imported them to the United States. It has been reported that four Furies remain in the Baghdad area.

Rarities in Fresh Water Lakes

From the harsh dry desert where Kittyhawk ET574 sat for seventy years, to cool, fresh water lakes half a world away, a number of rare World War II aircraft have been recently recovered and are slated for restoration.

In Japan's Aomori Prefecture, at the northern tip of the island of Honshu, a Japanese Army Type 1 twin-engine advanced trainer, known in the west as a Tachikawa Ki-54 (Allied code name "Hickory"), was recovered from the bottom of a lake in September 2012. Located a thousand feet below the surface of Lake Towada, the Hickory was recovered by volunteers from the Aviation Association of Aomori Prefecture.

This Ki-54 was assigned to the 74th Corps, Army Aviation Squadron 38, and was flying from Noshiro airfield (Akita Prefecture) to Hachinohe airfield (Aomori Prefecture) on September 27, 1943. Engine problems forced the crew to ditch the plane in Lake Towada, with only one of the crew of four surviving.

The Lake Towada Ki-54 is the only complete example of its type to survive. The Ki-54 was recovered in three sections—nose, tail, wings and center section—and plans call for the twin-engine trainer to be restored and displayed at the Misawa Aviation and Science Museum in Misawa City, Aomori. Two incomplete Ki-54s survive, fuselages only, one at the Australian War Memorial in Canberra, the other at the Beijing Aviation Museum, China.

In Canadian waters, Royal Canadian Air Force Northrop Nomad, serial number 3521, was discovered in Lake Muskoka on July 26, 2011. Nomad 3521 mid-aired with Nomad 3512; 3512 crashed on land while 3521 dropped into the lake. The wreckage was discovered by volunteers from Project LAMP (Lost Airmen of Muskoka). They in turn gave the crash site location to government divers who examined the aircraft in an effort to discover and recover any human remains. Once the remains of the crew are given a proper burial, Project LAMP would like to bring the aircraft up to serve as a memorial. This recovery effort continues.

In the Murmansk area of northern Russia, an Ilyushin IL-2 Sturmovik ground attack fighter was recovered from a lake in

Overhead view of Douglas A-20 Havoc, serial number 39-0741, sitting on the tundra more than eighty miles from Goose Bay, Labrador, Canada. *Underwater Admiralty Sciences*

September 2012. During a November 25, 1943, attack by sixteen IL-2s from the 46th Air Assault Regiment on the German airbase at Luostari, near the Norway/Russia border, the Sturmoviks were met by twenty-five Bf-109s. In the brutal air battle that ensued, eleven IL-2s were downed for a loss of twenty-three 109s. Pilot Valentine Skopintsev and Gunner Vladimir Humenny's IL-2's engine was knocked out

Bell helicopters from Canadian Forces 5 Wing, 444 Combat Squadron, used the Havoc recovery as a training exercise. They lifted sling loads of the A-20's parts to Goose Bay, Labrador. *Underwater Admiralty Sciences*

and the tail riddled with bullets during the engagement. Skopintsev belly-landed the fighter on a lake near Severomorsk, Murmansk, and although wounded, both air-crewmen survived. The Il-2 crew was credited with the destruction of two aircraft on the ground and gunner Humenny downed the 109 that was on their tail. Plans call for this aircraft to be combined with the wreckage from another IL-2 to make one flying Sturmovik.

On December 7, 2012, in the United States, the waters of Lake Michigan yielded another Grumman FM-2 Wildcat. This aircraft, BuNo 57039, was raised nearly sixty-eight years to the day of its December 28, 1944, crash. Ensign William Forbes was making his third take off to become qualified to fly from the decks of aircraft carriers when his engine quit. Momentum carried Forbes off the end of the USS *Sable*'s flight deck and into the water. Forbes landed in front of the carrier, which sliced the small fighter in two. Forbes was rescued, alive, albeit wet and bruised, and went on to become a naval aviator. The Wildcat's recovery was sponsored by aircraft collector Chuck Greenhill and brought to the surface by A. and T. Recovery. The fighter will be restored to static display standards by the Kalamazoo Air Zoo's Flight and Restoration Center in Kalamazoo, Michigan.

Discoveries at Sea

The sea has yielded a plentiful bounty of World War II aircraft recently. In late 2011, in U.S. waters off the coast of Jupiter, Florida, a dive charter operator discovered a Curtiss SB2C Helldiver in 185 feet of water. The charter operator turned the information over to the Naval History and Heritage Command in Washington, D.C., and in May 2013, divers from the Fleet Ocean Tug USNS *Apache* dove on the aircraft wreck. From photos, it appears this aircraft was stripped before being pushed over the side as there are no flaps, ailerons, or bomb bay doors, but the attach points can clearly be seen. The aircraft's identity is yet to be determined and it is likely that the dive-bomber will be "preserved in place."

In June 2012, OceanGate Inc. was testing its new *Antipodes* submersible off the coast of Miami Beach. *Antipodes* is fitted with new 2D and 3D high frequency sonar, and based upon a list of underwater targets provided by the National Oceanic and Atmospheric Administration (NOAA) were testing the submersible on what everyone thought was a sunken boat. OceanGate's high-frequency sonar showed that the target was a Grumman F6F Hellcat lying inverted on the ocean bottom with its engine barely attached. The as-yet-unidentified F6F was found in 240 feet of water, and is one of seventy-nine known Hellcat wrecks in the waters off Florida's east coast.

The following month, a NOAA team discovered the wreckage of a Brewster F2A-3 Buffalo in the lagoon of Midway Island. It appears the aircraft made a controlled landing on the water, and was discovered ten feet under the surface. This aircraft was ditched during a severe storm on February 12, 1942, four months before the Battle of Midway. Marine pilot Charles Somers Jr. survived the crash and later served with VMF-214, the famous Black Sheep squadron. He retired with the rank of lieutenant colonel and passed away in 1992. The wreck is a "micro-site" and will be documented but not recovered. It is the only documented wreck site of a Marine Corps Buffalo fighter.

Halfway around the globe, the wreckage of a six-engine Messerschmitt Me-323 *Gigant* (Giant) transport was located off the coast of Sardinia. The aircraft's 180-foot wing with its six engines lies two hundred feet under the surface. The *Gigant* was downed on July 26, 1943, by a Bristol Beaufighter on a flight between Sardinia and Pistoia, Italy.

In the seas of northern Europe, a Heinkel He-219 *Uhu* night fighter, a float-equipped Heinkel He-115 torpedo bomber, and a Ju-88 were found between April and June 2012. The He-115, Werke no. 2398, sank in 120 feet of water after a hard landing in the Hafrsfjord, near the island of Prestoy, north of Stavanger, Norway, in December 1943. In August 2012, the nearly intact torpedo bomber was lifted out of the water wearing the plainly visible fuselage code 8L+FH. The port engine was still on the wing, while the starboard engine was recovered

Robert Mester from Underwater Admiralty Sciences in the Seattle, Washington-area stands on the Havoc's wing and number one engine nacelle at the start of the recovery effort. Note that the engine is sunk halfway in the swampy terrain and that the wing has separated between the engine and the fuselage. *Underwater Admiralty Sciences*

Combat veteran C-47B, serial number 43-48716, *The Luck of the Irish* (seen on both sides of the nose), was hiding under a coat of white paint with a gold cheatline. This Skytrain flew with the 75th Troop Carrier Squadron supplying Allied troops during the Battle of the Bulge and later repatriated POWs from France back to England. After the war, the plane flew as a bug sprayer in Florida until acquired by the Vietnam Huey Pilots Association (VHPA). Air Heritage acquired the C-47B in a trade with the VHPA. *Courtesy Air Heritage, Inc.*

separately. The aircraft is now under restoration at the Flyhistorisk Museum Sola, in Sola, Norway.

At the beginning of summer 2013, the substantial remains of a Heinkel He-219 *Uhu* nightfighter came to the surface of Tannis Bay, between Hirtshals and Skagen, Denmark. The aircraft is now at the Garrison Museum in Aalborg, Denmark, where it will be restored. In the Baltic Sea, near Ruegen Island, major pieces of a Ju-88 were recovered for the German Historical Museum at Gatow Airport in Berlin. Only half the Ju-88 was recovered and plans call for a return trip to recover the balance of the bomber.

Volunteers at Air Heritage at the Beaver Falls, Pennsylvania, airport have stripped the white and gold paint from 43-48716 and are preparing to return the combat veteran to World War II configuration. *Tyler Pinkerton*

Found on the Tundra

A Lend-Lease C-47 was found in July 2012 on the Taymyr Peninsula in Siberia. The aircraft appears to have shut down the starboard engine in flight and was forced to land on the tundra. The port engine was making power as its propeller blades are bent and the engine sits awkwardly, having been torn from its mount. The Douglas-built C-47, serial number 42-32892 was delivered to the USAAF on February 23, 1943, and sent to the Russians on March 12. The plane was given the Russian serial number USSR-H-328 and was assigned to the 1st Air Division's 7th Arctic Aviation Regiment (AAR). As part of the 7th AAR, the C-47 flew ice reconnaissance patrols over the Kara Sea, above the Arctic Circle, mapping bergs and reporting that information to ships in the area.

The plane crashed on April 13, 1947, near the village of Volochanka. Eight crewmembers went for help, but were never seen again, while the remaining twenty-eight were rescued after twenty miserable days in the cold. Each year, the window for removing an aircraft from the Siberian tundra is extremely short; however, no plans have been announced regarding the aircraft's final fate. In many ways this C-47 is also an aviation time capsule as the majority of the instruments were in the panel and most of the interior fittings remain.

Robert "Bob" Mester and Mark Allen operate Underwater Admiralty Sciences, a nonprofit corporation, in Kirkland, Washington, where they have developed or improved upon numerous deep-sea diving and atmospheric diving suit (hard suit) technologies. They specialize in the exploration, identification, documentation, and recovery of submerged cultural resources. Thus they were selected to participate in the successful recovery of Vega-built B-17G 44-85740 from Dyke Lake in Labrador, Canada (see pages 64–87).

Mester and Allen learned of the substantial wreckage of a Douglas A-20 Havoc modified as an F-3A photo reconnaissance version, serial number 39-0741, sitting on the tundra in Labrador. The plane had been belly-landed on October 10, 1942, after engine troubles. The crew survived and was recovered, and the wreckage sat virtually untouched since the end of World War II, as it was located 85 miles from Goose Bay and 120 miles from the nearest road. During the course of four years, Mester and Allen dismantled the aircraft and, using commercial helicopters and those of Canadian Forces 444 Combat Squadron (5 Wing), shuttled the dismantled attack bomber to Goose Bay. The Canadian Forces used the opportunity to transport the aircraft components as training exercises, in addition to a desire to see the Havoc restored one day.

Once the A-20 was completely removed, it was loaded onto trucks and delivered to GossHawk Unlimited in Casa Grande, Arizona. Dave Goss and his company are known for their warbird restorations, including the maintenance and restoration of the ex-Doug Champlin Collection Focke Wulf Fw-190D-13, Yellow 10. The A-20 is in storage while Mester and Allen look for a sponsor or new owner to help fund the restoration of what will become the fourteenth Havoc survivor.

Hidden under a Coat of Paint

The Air Heritage aviation museum at Beaver County Airport, Beaver Falls, Pennsylvania, has won the aviation equivalent of the history lottery. The museum's volunteers have rebuilt a number of aircraft and were in search of a C-47 to add to their collection. The Vietnam

Huey Pilots Association, located at the same airport, was looking for an OV-1 Mohawk for its collection and had recently acquired an ex-sprayer C-47 from the State of Florida. Air Heritage happened to own an OV-1 Mohawk. A deal was worked out and ownership of the two aircraft changed hands. When Air Heritage took possession of the C-47B, they knew it was N836M, USAAF serial 43-48716, and that it had an overall white paint scheme with gold cheatlines.

Air Heritage was soon to learn that C-47B 43-48716 was completed on September 13, 1944, and subsequently delivered to the USAAF. The transport flew across the Atlantic Ocean and missed Operation Market Garden (the airborne operation to secure bridges across the Meuse, Waal, and Lower Rhine Rivers) by ten days. However, there was plenty of combat ahead. Transferred to the 8th Air Force, and then to the 9th Air Force, the C-47B flew with the 53rd Troop Carrier Wing, 435th Troop Carrier Group's 75th Troop Carrier Squadron. The plane flew two resupply missions during the Battle of the Bulge (December 16, 1944–January 25, 1945), and towed two Waco CG-4A gliders during Operation Varsity on March 24, 1945. At the end of the war, 43-48716 repatriated Allied POWs from France back to England.

Through the online warbird community known as "WIX"—Warbird Information eXchange—French researcher Patrick Elie provided the name of the C-47's pilot while operated by the 75th Troop Carrier Squadron. Air Heritage volunteer Tyler Pinkerton followed up with the pilot, Capt. Edward W. Frome, and his daughter to learn the markings of the aircraft. It wore nose art and the name *The Luck of the Irish*.

The aircraft's restoration is progressing nicely with the paint stripped and ready for application of its wartime colors. The interior of the C-47 will then be fitted out to World War II standards. Buying an unidentified C-47 that turns out to have extensive combat history is truly *The Luck of the Irish*.

Doing Good

Lou Sapienza, who was a team member of the P-38 *Glacier Girl* recovery expeditions, has set up his own exploration company, North South Polar, Inc., with the intent of locating and recovering missing-in-action and lost airmen. Most recently, Sapienza's efforts to find the missing U.S. Coast Guard Grumman J2F-4 Duck, serial number V-1640, from the cutter *Northland* lost in Greenland was the subject

A team from North South Polar and the U.S. Coast Guard has located the crash site of Lt. John A. Pritchard Jr., pilot, and Petty Officer 1st Class Benjamin A. Bottoms, radioman, seen here taxiing away from the cutter Northland. Pritchard and Bottoms were ferrying crewmen from a downed B-17 to the cutter when a storm closed in and the J2F-4 they were flying struck the terrain. Over the years, the Duck's wreckage became buried under thirty-eight feet of ice. *U.S. Coast Guard*

of Mitchell Zuckoff's 2013 book *Frozen in Time*. In January 2013, researchers from North South Polar and the U.S. Coast Guard located the Duck under thirty-eight feet of ice near Koge Bay, Greenland.

On November 29, 1942, U.S. Coast Guard Lt. John A. Pritchard Jr., pilot, and Petty Officer 1st Class Benjamin A. Bottoms, radioman, had departed the cutter *Northland* to continue rescue work bringing out the crew of a downed B-17E (serial number 42-5088) that had crashed on Greenland on November 9 of that year. Pritchard and Bottoms had landed on the ice and rescued two fliers earlier in the day, and that afternoon returned to bring out a third man, Army Air Forces radio operator Cpl. Loren Howarth, from the B-17E crash site. Using just the Duck's hull and outrigger pontoons to land and take off from the ice, the three men lifted off the ice cap for the eighteen-minute flight to the *Northland*, anchored in Comanche Bay (about twenty miles east by northeast from the bomber's crash site). Fog and

Surplus Australian Spitfires sit at the No. 6 Air Depot at Oakey air field, east of Brisbane after the war. Rumor has it that one or more Spitfires were removed from the storage yard, dismantled, and hidden on farms or in mine shafts. A group of enthusiasts has been following up leads for thirteen years, and they now plan to mount a full-scale effort to chase down those rumors. In the process, they plan to make a documentary film, titled *Broken Wings*, and have started a crowd-funding campaign to finance their efforts. The film's trailer shows a number of veterans; some say the aircraft were removed while others adamantly deny it. It will be a very interesting film. *Broken Wings*

storm clouds engulfed the area as the Duck lifted off and Pritchard became disoriented, flying into high terrain. In the ensuing years during the war and beyond, the Duck, its crew, and passenger became entombed in the ice.

When they returned from Greenland early in 2013, the North South Polar and Coast Guard crews presented their evidence of the Duck and its crew's final resting place to the service's command. A recovery attempt was authorized for August and September 2013, and although progress was made, the weather closed in early and the dig was halted for the year. North South Polar is slated to continue the recovery work in 2014. Once Pritchard and Bottoms are laid to rest, all missing Coast Guard personnel from all wars will have been accounted for.

It should be noted that Sapienza's experiences on the P-38 *Glacier Girl* warbird recovery were where he learned the techniques and gained the enthusiasm and desire to recover lost aircraft. In a way, warbird recovery can be cited as a driving force for bringing home these lost aviators.

Chasing Legends

The December 2012 expedition to uncover the buried Spitfires in Burma did not succeed for many reasons. However, money wasn't one of them. English farmer and aviation enthusiast David Cundall had been chasing the rumor that somewhere between 36 to 120 still-in-the-crate, Mk XIV Spitfires had been buried around air bases in Burma (today's Myanmar) at the end of the war.

Having spent more than £170,000 ($270,000) of his own money during the seventeen-year quest, plus approximately £1 million ($1.6 million) put up by sponsor Wargaming.net, the expedition fell apart in January 2013 when the excavation turned up electrical lines and water pipes feeding the Rangoon International Airport (the former RAF base Mingaladon) where the buried Spitfires were supposed to be located. After exploring sites at the former RAF base and at Myitkyina, archaeologists on the team concluded there are no Spitfires buried at either location. Cundall, however, believes the aircraft are there and pledged that the hunt would continue.

Farther south, in Australia, the rumor has persisted that Spitfires were removed from storage at the end of World War II and hidden in mineshafts and coal mines or buried in the vicinity of the Oakey air field. Oakey, which is just east of Toowomba and Brisbane,

Focke-Wolf Fw-190A-5, Werke no. 1501227, was hit by anti-aircraft fire near Leningrad and crashed in a swampy area, bordered by a forest. In the ensuing years, the trees grew up around the Focke-Wolf, hiding the fighter. It was recovered in 1991 and acquired by the Flying Heritage Collection in 1999. *Flying Heritage Collection*

was the home to No. 6 Air Depot (6AD), and shortly after the war ended, 554 aircraft—Boomerangs, Kittyhawks, Mustangs, and Spitfires—were stored at the field.

A documentary film project, known as *Broken Wings*, is trying to find a Spitfire, and has been following up on leads for the past thirteen years (www.brokenwings.com.au). They've gone to the crowd-funding source Pozible.com to solicit donations for the expedition and production of the film, and have begun to receive donations. Should a Spitfire or substantial remains of one be found, the Broken Wings team plans to donate it to The Museum of Australian Army Flying (located at Oakey) so that the general public will be able to see the aircraft of legend.

The *Broken Wings* film's website presents interviews with first-hand participants; some say that the aircraft were indeed transported off-site and "hidden," while others who were stationed at 6AD at the time believe, conclusively, that no aircraft were ever buried or stuffed into mine shafts and they had to account for every aircraft part and rivet. The producers say that either way—whether they find Spitfires or not—they'll have a real adventure proving or disproving the legend and the documentary will take viewers along on the exploration.

Fw-190s in the News

Wolfgang Falch of Sandy Air Corp. in Austria, has rebuilt Focke-Wolf Fw-190D-9, Werke no. 400616, which was last flown by *Unteroffizier* Koch with JG 54 Greenhearts. The fighter is identified by the number 16 painted in white on the aft fuselage. "White 16" is fitted with its original Jumo 213A and is composed of approximately fifty percent original parts, with the balance being new-build, and has been restored to static display standards. This doubles the population of long-nose 190s, the other being the ex-Doug Champlin/Flying Heritage Collection Fw-190D-13, Yellow 10, on display at the Museum of Flight in Seattle, Washington.

In addition to the Fw-190D-13, the Flying Heritage Collection of Everett, Washington, also owns an Fw-190A-5, a combat veteran from the Eastern Front. This aircraft was delivered in May 1943 and flew with 4./JG 54 (*Jagdgeschwader* 54—Fighter Wing 54; *Staffel* 4—Squadron 4) from Siverskiy, near Leningrad. At some point this aircraft was hit by anti-aircraft fire and force-landed in a swamp that bordered a forested area. The trees grew up around the aircraft, hiding it from view as it faded from memory of the locals.

Flying Heritage Collection's Fw-190A-5 is the only one of its type flying with an original BMW 801 radial engine. Steve Hinton flew this aircraft for its first post-restoration flight on December 1, 2010. *Jim Larsen via Flying Heritage Collection*

In 1991, the aircraft was recovered and transported to Personal Plane Services in England. Warbird collector Doug Arnold acquired the aircraft in 1991, and the plane passed to the Flying Heritage Collection in Everett, Washington, on February 1, 1999. After a lengthy restoration, the Werke no. 400616 was delivered to the Flying Heritage Collection's facility and today is the only Fw-190A flying with its original BMW 801 radial engine. The use of the original BMW 801 engine on the Fw-190 restoration enables enthusiasts to see and hear this aircraft in its element, something not possible for the past sixty-nine years.

What does the future hold for warbird recoveries? The possibilities are endless as new areas of World War II's battlefields are opened and explored. One thing is certain, they're still out there . . .

Part One

What Historic Aircraft Do the Waters Hold?

Lakes, rivers, and the ocean hold a time capsule of World War II military aviation. Allied and Axis aircraft alike sit on the bottom, or appear and disappear with the shifting sands. A number of once-extinct aircraft have been brought to the surface recently, and other historic aircraft are awaiting their turn.

One historic fighter that was recovered from the sea has recently flown in the United Kingdom. In September 1980, a Mark I Spitfire (RAF serial P9374) that went down during the Battle of France was recovered from the shifting sands off Calais by the Musee de l'Air at Le Bourget Airport. Although it wasn't much to look at, it formed the basis of a rebuildable project—for someone with the time, technical skill, and financial means to make it happen. This early-mark Spitfire was painstakingly rebuilt and flew for the first time on September 1, 2011, with 2012 being its first full year of flying since 1940. When it took to the air on that early fall day, P9374 became the oldest Spitfire flying, and one with a confirmed combat history.

Half a world away, the Pacific Ocean holds a treasure trove of World War II aircraft. When the USS *Yorktown* was sunk at the Battle of Midway on June 7, 1942, a number of Douglas TBD Devastator torpedo bombers and Grumman F4F Wildcats lashed down on her hangar deck went to the bottom. Before the ship was sunk by a torpedo from Japanese submarine I-168, those killed in attacks on the carrier were buried at sea, making this ship an artifact of war, rather than a war grave. More than fifty years later, on May 19, 1998, Dr. Robert Ballard found the ship sitting upright, with a twenty-five

Thousands of aircraft ended up sinking to the bottom of lakes, rivers, and the oceans of the world during the war. Salt water is very harsh on aluminum and magnesium, however, those found in fresh water lakes have great potential to be returned to the air. *Courtesy Alan Griffith*

degree list to starboard in 16,650 feet of water. The ship appears to be preserved in excellent condition, and one has to wonder if there will come a day when technology, the rights to recover and restore the aircraft, and the money needed to accomplish this will come together to see a number of combat veteran Wildcats and Devastators break to the surface. Only time will tell.

In addition to the tantalizing finds above, the following five chapters examine other Allied and Axis fighters and bombers, each recovered from salt or fresh water. Each aircraft represents an historic era of World War II aviation.

Battle of Britain
Do-17 Recovery

Many good sea tales begin with a fisherman snagging his net on an underwater object, and this tale is no different.

In 2000 or 2001, a fisherman snagged his net while fishing the waters of Goodwin Sands, fronting the White Cliffs of Dover near the town of Deal in Kent, England. Dover sits at the English Channel's narrowest point, and faces Calais, France. This area took the brunt of the fighting during the Battle of Britain (July 10–September 6, 1940) and the ensuing Blitz on London and surrounding cities (September 7, 1940–May 21, 1941). Kent was known as "Hell Fire Corner" for the amount of destruction that rained down on the area during the battles.

Rare color image of a Do-17Z in flight. More than 1,500 examples of the "Flying Pencil" were built, and more than 400 attacked England during the Battle of Britain and the subsequent Blitz on London. The Do-17Z-2 (fuselage code 5K+AR) was shot down on August 26, 1940, and ditched off the coast of Dover. *Courtesy RAF Museum*

Estimates put the number of shipwrecks in this area of the English Channel at more than two thousand. To illustrate the point, the area was the site of two major naval battles during World War I: the First Battle of Dover Strait (October 26–27, 1916, when German forces were victorious over the British), and the Second Battle of Dover Strait (April 20–21, 1917, a British victory).

Because of the area's huge number of sunken ships, most in the fishing community were not surprised when the nets were snagged and subsequently repaired, and the incident was soon forgotten. In 2004, sport scuba diver Bob Peacock learned of the wreck's location, but it was another

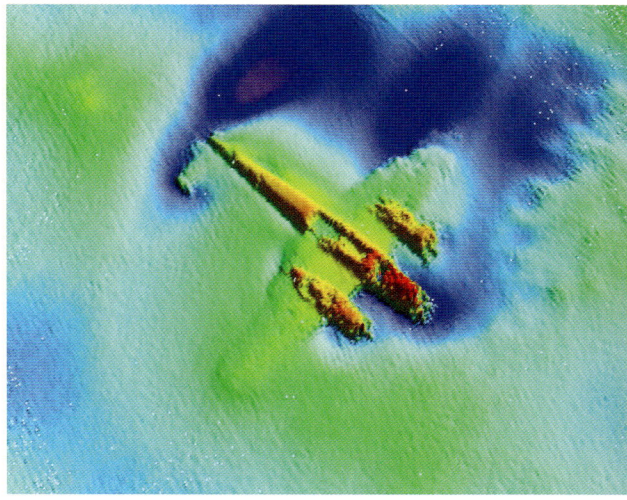

Side-scan sonar image of the Do-17Z prior to recovery by the Royal Air Force Museum. The scan shows the Do-17 resting inverted on the bottom of the English Channel. The aircraft was ditched into the sea, and then turned upside down during its descent to the bottom of the channel. From this scan, it can easily be determined that the starboard horizontal tail and the bomb bay doors are missing. *Port of London Authority*

four years, until September 2008, before he made the journey to look for the snag. Lying almost four miles off the coast, the bottom topology is a chalk bed, between fifty and eighty feet deep covered by shifting sands. Underwater visibility in the area ranges from twenty-five feet on the best day, to nearly zero when the tides are running.

Instead of a shipwreck, Peacock found an aircraft partially buried in the sand. Peacock reported his find, and Wessex Archaeology of Salisbury, England, was given the task of surveying the crash site. In May and June 2009, Wessex Archaeology found a twin-engine aircraft lying inverted, nearly complete, fifty feet below the surface. Around the aircraft was a debris field, but a number of parts were missing, including the starboard horizontal and vertical tail, tail cone and tailwheel, flaps, engine cowlings, bomb bay doors, main landing gear doors, and the cockpit canopy.

From a small number of parts recovered from the wreck in 2009, researchers from the RAF Museum, Hendon, England, were able to determine that the wreck is that of a Dornier Do-17-Z2—the

Above and opposite: On June 11, 2013, the Do-17Z was raised from the sea in an area known as Goodwin Sands, approximately four miles from the English coast. Divers attached lifting straps to the aircraft and it was raised from the bottom and lifted onto a barge. *Courtesy RAF Museum*

only substantially complete example of its type known to exist. The Do-17Z was a light, fast bomber, referred to as the *Fliegender Bleistift* or "Flying Pencil." There were more than 1,500 Do-17s built between 1934 and 1944, and more than 400 were used against the British during the Battle of Britain. The Germans lost more than 200 Do-17s to all causes during the Battle of Britain and subsequent Blitz.

RAF Museum researcher Andrew Simpson compiled the Do-17Z's history, believing the aircraft to be Werke no. 1160 that was based with 7 *Staffel*, III/KG3 (7 *Staffel* or Squadron, 3rd Group, *Kampfgeschwader* or Bomber Wing 3). At the beginning of the Battle of Britain, KG 3 had 108 bombers on strength, of which 88 were combat ready. The unit was based at St. Trond, Belgium, and on August 26, 1940, Werke

no. 1160, wearing the fuselage codes 5K+AR, was part of a squadron sent to bomb the RAF airfields at Debden and Hornchurch, more than sixty miles north and east of London. Flying above clouds, the Do-17 became separated from its squadron and turned south to get its bearings. The Do-17 was attacked by Boulton Paul Defiants—low-wing, turret-equipped fighters—of No. 264 Squadron, based at RAF Hornchurch. Although accounts are conflicting, in the ensuing melee, seven Do-17s of the attack were shot down by the Defiants.

It is assumed that Do-17 5K+AR made a run for the coast headed for Calais, France, because its crash location is some eighty miles south and east of the Germans' intended target. A direct line back to St. Trond required a flight of more than two hundred miles, while

Marine growth covers every square inch of the German bomber. In spite of how bad it looks in this view, the growth helped preserve many parts. *Courtesy RAF Museum*

Calais, or the narrow channel area between Dover and Calais, afforded a better chance of survival for a bomber trying to limp home, short of fuel. If the Do-17's crew had to ditch, they stood a greater chance of being picked up by German crash boats patrolling in the vicinity.

Hit by fire from the RAF Defiants, 5K+AR ditched in the Channel, less than four miles off the English coast, not far enough across to hope for rescue by the Germans. Pilot Feldwebel Willi Effmert, and bombardier *Unteroffizier* Hermann Ritzel were both wounded in the air battle. After the aircraft ditched, they were rescued by the British and became prisoners-of-war in Canada. Wireless operator *Uffz.* Helmut Reinhardt and bombardier Gefreiter Heinz Huhn were killed. Their bodies subsequently washed ashore on opposite sides of the Channel with Reinhardt interred in Holland and Huhn buried in England. While the pilot and bombardier were being rescued, Do-17Z-2 5K+AR slid beneath the English Channel, landing inverted on the bottom.

Recovery Operation

In June 2010, another series of dives were made to access the condition of the wreck and to plan for the Do-17Z's recovery. The plan involved lifting the entire wreck in one piece through a cage system. Having determined a positive potential for recovery, the project to recover the sole surviving Flying Pencil began to move forward.

"The discovery and recovery of the Dornier is of national and international importance. The aircraft is a unique and unprecedented survivor from the Battle of Britain and the Blitz," said Air Vice-Marshal Peter Dye, director general of the RAF Museum.

The recovery operation received more than £345,000 ($538,000) from Britain's National Heritage Memorial Fund and the project got underway. The RAF Museum served as manager of the recovery project with assistance from Seatech Commercial Diving Services, the Port of London Authority, and researchers from Imperial College, London, who are serving as conservation consultants. Additional funding was provided by WarGaming.net, 328 Support Services GmbH, the EADS aerospace consortium (of which Dornier is a legacy company), the Society of Friends of the RAF Museum, and the general public.

With everything in place, the RAF Museum and Seatech team began operations in May 2013. Poor weather in the Channel forced the team to abort recovery attempts and return to port four times. With so many aborts, the RAF Museum decided to abandon the cage recovery system in favor of attaching lifting rigs directly to the aircraft. Divers worked in teams, and were only able to stay down for forty minutes at a time rigging the new lift system. The new rigging would enable a seagoing crane to make one lift of the bomber, from the bottom of the sea up and onto the deck of a barge.

On the eve of the lift, June 2, 2013, RAF Museum Director Dye said, "We have adapted the lifting frame design to minimize the loads on the airframe during the lift while allowing the recovery to occur within the limited time remaining. The RAF Museum has worked extremely close with Seatech throughout this process and both organizations remain determined to complete this challenging task and see the Dornier safely recovered as planned." Unfortunately, the winds did not favor the recovery attempt on June 2, and the crew watched the weather, waiting for a new opportunity.

On June 10, the winds died down, the seas calmed, and the RAF Museum and Seatech team was able to make a successful lift. After a long day of dedicated work, the Do-17Z was brought up on deck at

Manufacturer's dataplate as found on the Do-17Z, seen shortly after recovery. *Courtesy RAF Museum*

6:30 p.m. local time. The main landing gear tires were still inflated and the propellers, recovered separately, were curved, showing that they were turning under power when the bomber ditched.

Once on deck, any hazardous materials were addressed and the Do-17's MG 15 machine guns and magazines were secured for later inspection. When safely back on the dock, the magazines were x-rayed, only to be found empty—a vivid reminder of the back-and-forth air battle fought in the skies over Kent.

During the five-hour trip from the recovery site to the dock at Ramsgate Harbor, the RAF Museum team went to work removing the bomber's wings and horizontal tail. After they had cleared seventy-three years of marine growth, the attach bolts were easily removed. The fuselage, wings, and tail were put into a frame for the loading and transport of the German bomber to the RAF Museum at Cosford for conservation. As the frame was being built around the bomber's components, a gel was sprayed and brushed onto the aircraft to keep oxygen from continuing to corrode the metal.

The aircraft was driven more than two hundred miles on flatbed trucks to the Michael Beetham Conservation Center at Cosford. Once at the museum, the aircraft components were unloaded and placed into two hydration tunnels, essentially large greenhouse-like structures, where they are repeatedly sprayed with citric acid as the first step in conserving the aircraft. In addition to inhibiting corrosion, the citric acid helps soften the layers of marine growth, enabling further corrosion-inhibiting efforts. As the marine growth becomes loose, conservation staff uses plastic scrapers to peel back the sea slime.

The conservation process is expected to take two years. To engage guests to the RAF Cosford Museum while the aircraft is conserved, a visitors' center and an education center will enable the Dornier to be seen up close while in the hydration tunnels. An RAF Museum spokesman said that displays "at the Cosford and Hendon Sites will 'taxi' the aircraft virtually through the wall and at the Hendon site the aircraft will be raised through the 'sea' of carpet. The nose windows

Above: The fuselage was lifted from the sea in its inverted position, and was transported in the same orientation. *Courtesy RAF Museum*

Right: Dornier Do-17Z fuselage being placed into the hydration tunnels at the Michael Beetham Conservation Center at the RAF Museum, Cosford. The aircraft is sprayed with a citric acid to inhibit corrosion. The preservation process is expected to take more than two years. *Courtesy RAF Museum*

will light up to form a projection-mapped screen and tell the story from the gunner's seat. During this phase virtual Dorniers will appear and hover ominously above their external shadows. Visitors will be able to view and explore these augmented-reality, life-size 3D models though a mobile application on their smart phones."

The RAF Museum is no stranger to recovering and preserving aircraft, having most notably recovered a Halifax Mk II in 1973 from Lake Hoklingen in Norway and a Hurricane Mk I retrieved from the Thames estuary. It is conceivable that the Do-17, when ready for display, will sit facing the Thames estuary-recovered Hurricane in a Battle of Britain display.

The Dornier Do-17Z "will provide an evocative and moving exhibit that will allow the museum to present the wider story of the Battle of Britain and highlight the sacrifices made by the young men of both air forces and from many nations. It is a project that has reconciliation and remembrance at its heart," said RAF Museum Director General Dye.

Bringing Up a Birdcage Corsair

Vought Aircraft's prototype F4U Corsair flew its maiden flight on May 29, 1940, with company test pilot Lyman Bullard at the controls. The navy fighter featured a 13-foot, 4-inch diameter propeller, which required designers to incorporate "gull wings." The plane was very fast and on October 1, 1940, became the first single-engine U.S. military aircraft to fly straight and level at more than 400 mph (the twin-engine Lockhheed P-38 flew faster than 400 mph in January 1939).

Although fast and packing a heavy punch with six .50-caliber wing-mounted machine guns, the Corsair suffered from some serious teething troubles. On September 25, 1942, F4U-1 Bureau of

F4U-2 BuNo 02155 sits on the ramp at Vought's Stratford, Connecticut, factory. This Corsair is from the same batch as BuNo 02465, which later sank in Lake Michigan. The cockpit canopy with multiple bracing struts and the raised center panel for a rear-view mirror earned this Corsair model the nickname "Birdcage Corsair." *Vought Archives*

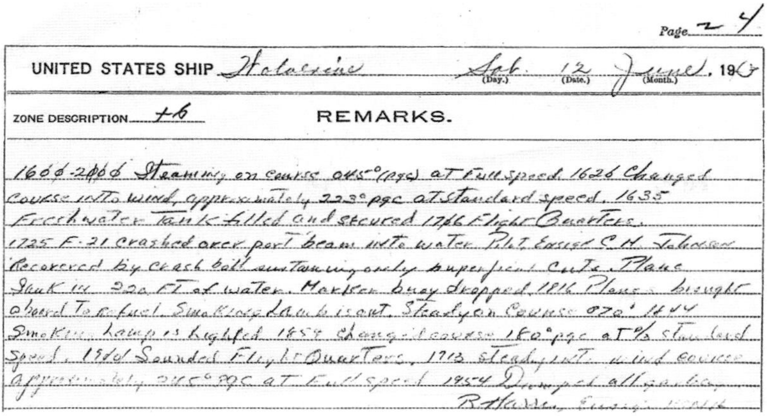

UNITED STATES SHIP _Wolverine_ ____ _Sat._ _12_ _June_, 19_43_
(Day) (Date) (Month)

ZONE DESCRIPTION ___+6___   REMARKS.

[handwritten log entry]

1600-2400 Steaming on course 045° (pgc) at Full Speed. 1626 Changed course into wind, approximately 223° pgc at Standard speed. 1635 Freshwater tank filled and secured 1716 Flight Quarters. 1725 F-21 crashed over port beam into water. Pilot Ens. C.H. Johnson Recovered by crash boat sustaining only superficial cuts. Plane sank in 220 ft. of water. Marker buoy dropped 1816 Plane brought aboard. Took fuel. Smoke screen is out. Steady on Course 070' 1844 Smoke lamp is lighted 1859 changed course 180° pgc at ½ standard speed. 1900 Sounded Flight Quarters. 1913 steady, into wind course approximately 045° pgc at Full speed 1954 Damped all [illegible].

R. Harris Ensign USNR

When A. and T. Recovery went looking for the Corsair, the only resource they had was this log entry from USS *Wolverine* for June 12, 1943, which reads: "1725. F-21 crashed over port beam into water. Pilot Ens. C.H. Johnson recovered by crash boat sustaining only superficial cuts. Plane sunk in 220 feet of water. Marker buoy dropped." Not much to go on, but having mapped the southern end of Lake Michigan, it was only a matter of selecting targets 220 feet below the surface and eliminating them until they came to the Corsair. *A. and T. Recovery*

Aeronautics serial number (BuNo) 02156 was flown aboard USS *Sangamon* in Chesapeake Bay for carrier qualifications. During the plane's course of four takeoffs and four landings aboard ship, the engine coated the windscreen with a fine film of oil; it was difficult for the pilot to see the landing signal officer (LSO) when in the three-point landing attitude; the fighter's stiff landing gear caused the plane to bounce on landing; and the left wing had a tendency to stall before the right wing, especially during deceleration.

To address these issues, Vought sealed the top three cowl flaps and replaced the individual hydraulic cowl flap actuators with a single hydraulic cowl flap master actuator and mechanical linkage to the remaining cowl flaps. The wing stall issue was solved by adding a small wooden spoiler on the right wing, causing both wings to stall at the same time. The cost of adding the wing stall device was $3.67. The first F4U-1 to receive this modification was BuNo 02510.

It took a year to improve the aircraft's landing characteristics and simultaneously address the pilot vision problem. To reduce the landing bounce, Vought replaced the main gear oleo's Schrader valve with Chance Vought valve number 44213 and the strut's air pressure was increased. This change was made on the production line, and all previously built Corsairs were modified during major overhauls.

The tail wheel yoke was redesigned and extended by six inches, which improved the pilot's forward vision on the ground and changed

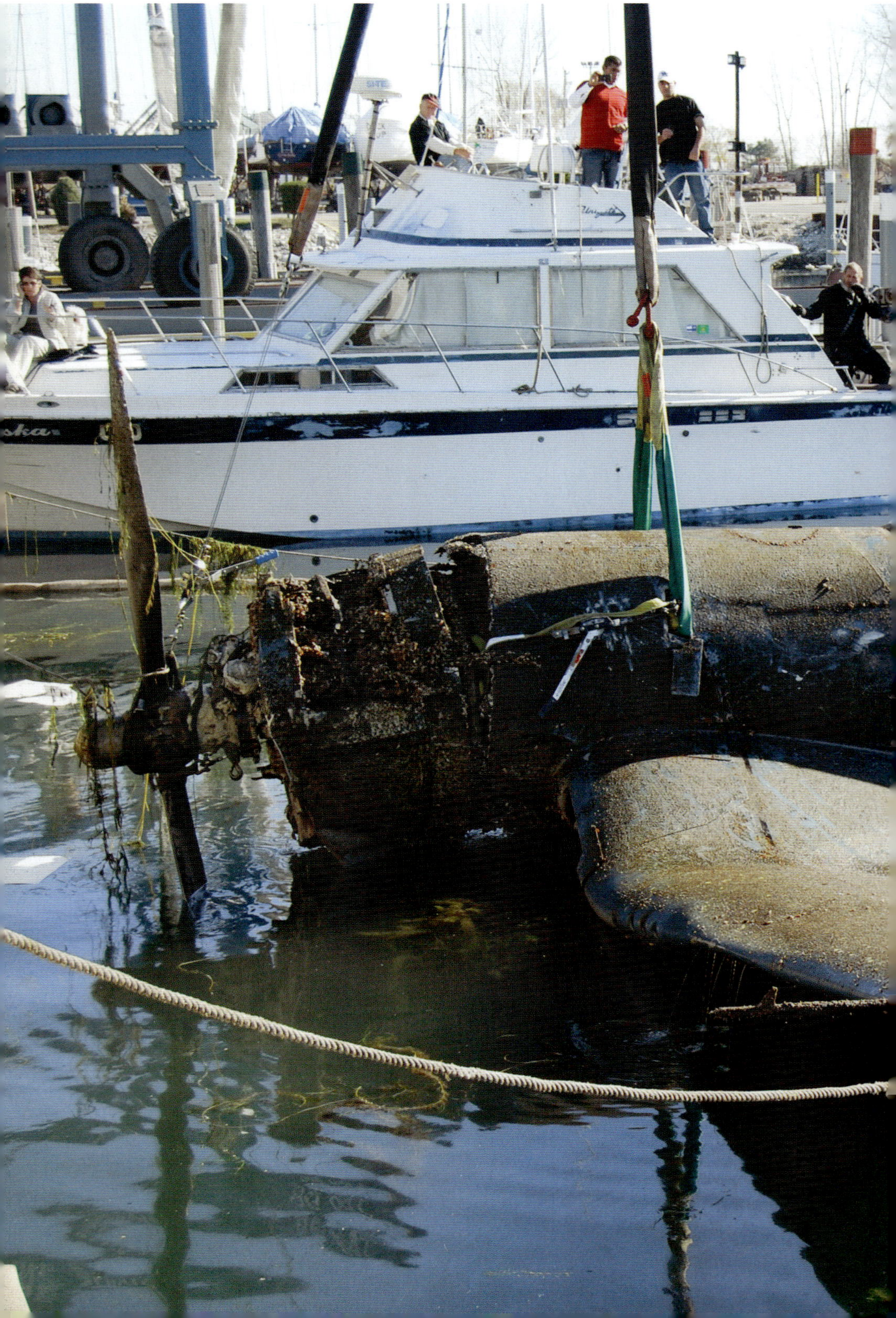

the tail hook down-angle from seventy-five to sixty-five degrees. This change helped prevent Corsairs from "sitting" on their tail hooks during full stall landings. The 404th Corsair, F4U-1 BuNo 02557, was the first aircraft on the assembly line to incorporate this change. It should be noted that the changes made to the landing gear oleos and change resulting from the raised tail wheel yoke reduced the Corsair's takeoff distance by twenty feet in a twenty-five knot head wind.

While the shipboard issues were worked out, Corsairs joined land-based marine fighter squadrons in late 1942, VMF-124 being the first to receive F4U-1s in October. The squadron arrived on Guadalcanal on February 12, 1943. In November 1943, Solomon Islands land-based VF-17 began taking its Corsairs aboard ship during raids on Rabaul and other Japanese strongholds. The VF-17 Corsairs would land aboard ship, refuel, then continue home.

In all, Vought, Brewster, and Goodyear Aircraft built 12,571 Corsairs of all models, and during World War II, the type achieved an 11:1 kill ratio over the Japanese.

The 313th F4U-1, BuNo 02465

On Saturday, June 12, 1943, Vought's Corsair was still not ready for shipboard service. That didn't preclude Ensign Carl H. Johnson (U.S. Naval Reserve–USNR) from being sent out to qualify aboard ship in F4U-1 BuNo 02465. Johnson had accumulated 388.5 hours of flight time and 57 hours in the Corsair.

BuNo 02465 wore the fuselage code F-21, and was virtually factory-fresh. It is known as a "birdcage" Corsair: the reinforcing bars on the pilot's canopy make the pilot look like a bird in a cage, hence the name. BuNo 02465 was accepted by the Navy on March 26, 1943, at NAS New York, and three days later was delivered to the Carrier Qualification Training Unit at NAS Glenview, Illinois.

At 5:06 p.m., local time, the training carrier USS *Wolverine* (IX81) turned into the wind on Lake Michigan to begin flight operations. The weather was good, and around 5:20 p.m., Ensign Johnson entered *Wolverine's* landing pattern. As he approached the stern of the carrier, Johnson lost sight of the LSO, the view blocked by the Corsair's long nose. Missing the cut signal, Johnson applied power to go around

Knowing that the National Naval Aviation Museum did not have an early model Corsair, warbird collector Chuck Greenhill generously sponsored the recovery costs of BuNo 02465. Before the rare Navy fighter had been transported to the museum's restoration shops in Pensacola, Greenhill was already making parts for the Corsair's restoration. *A. and T. Recovery*

again. Unable to apply enough power to stop the Corsair's sink rate, the 11,142-pound fighter continued to settle toward the carrier's deck.

As the Corsair mushed to the deck, its tail hook caught the number six wire, breaking the wire, and then snagging the number seven wire. The number seven wire stretched to its stops and as the battle of Corsair versus the last arresting wire played out, the wire won. The strength of the wire stripped the tail hook assembly out of the Corsair's fuselage and the resulting change in momentum pointed the aircraft down and over the port side. Johnson rode the fighter into the water and it floated long enough for him to get unstrapped, out of the cockpit, and into the lake. He was soon picked by the crash boat and only suffered minor cuts and bruises.

BuNo 02465 was laid out as to look like a whole airplane for a series of progression photos. Note that the fuselage has been separated at the production break behind the cockpit and that the horizontal tail has been removed from the fuselage. *National Naval Aviation Museum, Paul Schumacher*

On the bridge of *Wolverine*, officer of the deck Ensign R. Harris (USNR) wrote, "Plane sunk in 220 feet of water. Marker buoy dropped." He marked the time as 5:25 p.m.

On June 16, 1943, Ensign Johnson returned to the decks of *Wolverine* at the controls of SNJ-4C Buno 05621. He successfully completed his series of four takeoffs and landings in this aircraft, qualifying him as a naval aviator. He headed west to Hawaii and eventually was posted to Fighting Ten (VF-10) Grim Reapers.

Under the command of James H. Flatley, VF-10 took its Grumman F4F Wildcats into combat aboard USS *Enterprise* (CV-6) during the Battle of the Coral Sea and then to the Battle of Guadalcanal. The squadron had returned to the U.S. West Coast and traded in its

Wildcats for the new Grumman F6F-3 Hellcat. As the squadron was working up at NAS Puunene, Maui, Territory of Hawaii, Ensign Johnson joined its ranks as a Hellcat pilot.

Unfortunately for Johnson, he would never see combat. On November 25, 1943, he was involved in a midair collision with Ensign Eugene Sumner (USNR) during a training flight. Sumner was able to parachute from his F6F-3, BuNo 40037. He broke his shoulder and collarbone during the jump, and his Hellcat was completely destroyed. Johnson was at the controls of F6F-3 BuNo 40038, and he did not survive the midair contact, perishing in the crash.

Fast-Forward Fifty Years

In the early 1990s, Allan Olson and Taras Lyssenko formed A. and T. Recovery in the Chicago area. The two divers and explorers had obtained copies of the Great Lakes training aircraft carriers' logbooks and began to map the bottom of Lake Michigan in search of the more than

140 aircraft lost over the sides of USS *Sable* (IX64) and USS *Wolverine*.

The pair used information from the ship's logs as their starting point for side-scan sonar searches. Lyssenko and Olson mapped the southern end of Lake Michigan in the areas where the two training carriers operated most often. From the maps, they were able to build a database of likely targets. Using the target list, they began to videotape each one, building an incredible library of the Navy planes at the bottom of Lake Michigan.

Having developed a list of targets and locations, it was only a matter of matching what was found on the bottom to the list of types lost during wartime training. The numbers include forty-one TBM/TBF Avengers, one F4U Corsair, thirty-eight SBD Dauntless dive-bombers, four F6F Hellcats, seventeen SNJ Texans, two SB2U Vindicators, thirty-seven FM/F4F Wildcats, and three TDNs—early radio-controlled drones.

Mussels and other marine growth can be seen on the underside of the starboard wing prior to cleaning and disassembly. Notice the Corsair's engine air intake scoops forward of the firewall. *National Naval Aviation Museum, Paul Schumacher*

F4U-1A BuNo 02465 was found in the mid-1990s. On sonar, it didn't look like an aircraft, but a remotely operated vehicle (ROV) photo showed a Corsair on the bottom missing its tail. The fighter was sitting in a three-point position resting on its main landing gear, engine, and prop. Its aft fuselage was pointing toward the surface, 240 feet above.

A. and T. Recovery had hoped to recover the Corsair in exchange for another aircraft, or the rights to salvage an aircraft from Lake Michigan. Through a series of delays, the Corsair recovery, and that of a Hellcat (F6F-3 BuNo 25910), were put off until 2009. Warbird collector Chuck Greenhill agreed to foot the recovery bill for the

Stripped and mounted nose down to a fixture, disassembly of the Corsair proceeds in the museum's restoration center. The next step will see the wings separated from the fuselage. *National Naval Aviation Museum, Paul Schumacher*

F4U-1A and Andy Taylor from Enterprise Rent-A-Car funded the Hellcat's recovery. Taylor's father had flown Hellcats from the USS *Enterprise* and USS *Essex* during World War II, and the recovery and restoration of the fighter was a fitting tribute to Taylor's father, Jack C. Taylor. The Hellcat was brought up in November 2009, and then A. and T. Recovery began to concentrate on the Corsair.

A year after the Hellcat was raised, A. and T. Recovery set out to bring the F4U-1A to the surface. As Taras Lyssenko explained:

Diving conditions on Lake Michigan are pretty miserable in November. It's freezing cold. The visibility is terrible under water and you can only see ten to fifteen feet. You've got to be real careful as you don't want to lose a hundred-something thousand dollar piece of equipment.

With the aircraft nosed-over, we're very cautious. We don't want to go under the aircraft with the ROV, and when driving it, you don't have much peripheral vision. We only see what the ROV camera sees directly in front of it. Since the invasion of the zebra and quagga mussels, they've changed the visibility of the lake. The first thing we saw was the tail section, broken off and sitting to one side. Apparently it had been hanging by wires for several years and there it was underneath the aircraft. We could not see it in the 1990s, but it was quite visible in 2010. [*The detached tail section was the first piece recovered on November 2, 2010.*]

The crew on the boat is pretty much consistent almost all the time. It's about six, seven people. They go down and they attach lifting bags wherever I want them, whether it's fifty feet down or whatever. They dive down each time, install the bags and then when the aircraft gets up to less than a hundred feet they dive back down to make sure the aircraft is secured and tie on any loose pieces. We attached the lift bags every ten feet or so, and the bags are vertically linear. They slowly lift the aircraft off the bottom.

The lifting system is like a giant rubber band, and we can attach bags all along the length of the system however we rig it. Then the bags are inflated from the surface as needed.

With the Corsair floating just below the surface, A. and T. Recovery's boat began to tow the fighter to the Waukegan Harbor forty miles away. Waukegan Harbor has a large crane used to lift boats in and out of the water, and it is ideally located on the north side of Chicago—closer to where the aircraft wrecks are located.

The fighter was towed into the harbor submerged, then lifted out onto an

Manufacturer's dataplate for F4U-1 Corsair BuNo 02465 shown mounted in the cockpit before the aircraft was cleaned. *National Naval Aviation Museum, Paul Schumacher*

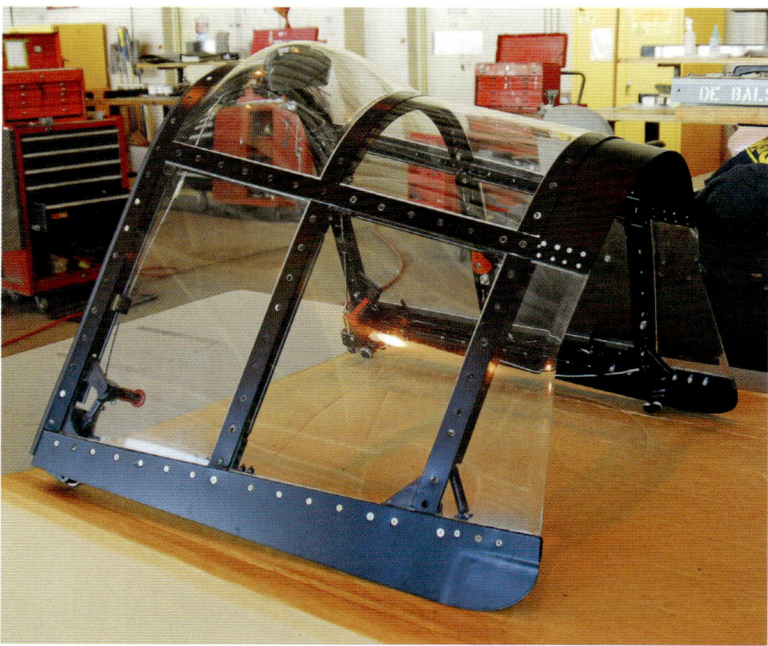

The birdcage canopy before and after restoration by the museum's restoration staff. *National Naval Aviation Museum, Paul Schumacher*

The port-side flap after repair and priming. This level of detail will ensure the Corsair is around for generations to see and learn its history and contribution to Naval aviation. *National Naval Aviation Museum, Paul Schumacher*

environmentally safe pad that prevents any fluids from the aircraft reentering the water. It also contains any oil or hydraulic fluid that might leak out when the plane is being lifted onto the dock. F4U-1A BuNo 02465 touched dry land on November 8, 2010.

A. and T. Recovery then prepared the aircraft for transportation to the National Naval Aviation Museum in Pensacola, Florida. Any mud, silt, and debris brought up with the aircraft stayed where it was and was transported to the museum where aircraft conservators cleaned and inspected the aircraft.

Today the F4U-1A is undergoing restoration at the Naval Museum's facility. Chuck Greenhill's generosity has extended to funding the restoration, and he is making some parts for the aircraft as well. When completed, BuNo 02465 will be one of seven surviving early F4U-1, -1A, FG-1, -1A, and F3A-1 Corsairs.

The windscreen and glare shield after the primer paint coat has been applied. Restoration will take a few more years. *National Naval Aviation Museum, Paul Schumacher*

Ju-88 U4+TK
Recovery and Restoration

Ju-88A U4+TK begins to sink under the ice of Jonsvatnet Lake. Notice that the engine cowlings have been removed and the emergency hatch on gunner's cupola hangs open. *John Njålsund collection via Guttorm Fjeldstad*

Carrying a maximum payload of 3,100 pounds, the Junkers Ju-88 twin-engine bomber was employed extensively during the opening stages of World War II as a bomber, dive and torpedo bomber, night fighter, and reconnaissance platform. In the closing months of the war it was also part of the Mistel weapon, an explosives-laden, remotely controlled flying bomb. In all, 15,183 Ju-88s were built in ten significant models with numerous variants.

Currently under restoration at the Norwegian Armed Forces Aircraft Collection, at the Gardermoen Air Force Museum, north of Oslo, Norway, is Junkers Ju-88A-1 Werke no. 0880119. This aircraft first flew on January 2, 1940, and the restoration team believes it may have been involved in an accident early in its career; as the team works to restore 0880119, they have found parts from 0880099, 0880114, 0880121 and other aircraft inside. The aircraft was repaired quickly because it then took part in Operation *Weserübung*, the Nazi invasion of Denmark and Norway on April 9, 1940. During this combat assignment, the bomber wore the fuselage code of U4+TK, with the "T" written in white to denote its assignment to Bomber Wing 30 (KG 30).

The Nazis' Operation *Weserübung* sought the Norwegian ports for use as future U-boat bases; their close proximity to the Allies' Atlantic Ocean shipping lanes; and to preserve their supply of iron ore from Sweden, the majority of which came through the northern Norway port of Narvik. In addition, Germany's control of these ports prevented the Allies from using them to block Nazi access from the Baltic Sea into the Atlantic Ocean.

During the battle, the German Army was capturing territory so quickly that the Luftwaffe was forced to improvise a number of open spaces for use as airports. When the Germans attacked Trondheim on April 9, they found the airfield at Vaernes unprepared to handle the

A pair of Luftwaffe mechanics check out the sinking U4+TK. The canopy cover has been removed as the bomber sinks. *Thomas Seidel via Guttorm Fjeldstad*

Prior to the invasion of Norway, this photo of a number of KG 100 Ju-88As line the ramp. Note the U4+ fuselage markings. *Courtesy Guttorm Fjeldstad*

weight of its transport planes and bombers. Until the runway could be improved, the Luftwaffe began using nearby frozen Jonsvatnet Lake as a forward-operating base. The lake, five miles south and east of Trondheim, opened as an airstrip on April 11, when eighteen Ju-52s arrived with ammunition. This was followed by the arrival of Ju-88s of KG 26 and KG 30.

On April 15, bombers from Jonsvatnet attacked the town of Namsos, burning it to the ground. And although Jonsvatnet Lake would only be used for two weeks, it provided a landing field for troop-laden Ju-52s that would form an attack force to move up the coast. Once the airfield at Vaernes was opened, the Germans were able to airlift in more than 1,500 troops a day. However, all of the aerial activity in the area attracted the attention of British reconnaissance aircraft that discovered the frozen landing ground on Jonsvatnet Lake.

On April 22, a Wellington bomber flew over Jonsvatnet Lake and reported more than fifteen aircraft operating from the improvised airfield. The following day, the Luftwaffe began moving its aircraft to Vaernes, where wooden planking had been laid down to improve the runway surface. Only an He-111 and a Ju-88 were left at the field, sinking through the ice. On April 25, RAF Skuas strafed the two aircraft, setting the wings of the remaining Ju-88, U4+TK, on fire.

At some point early in the invasion of Norway, U4+TK was heavily damaged in combat but was able to limp back to base. Here it was moved to one side and most likely became a source of spare parts for other KG 30 Ju-88s. When the time came, U4+TK was abandoned on the ice with the RAF Skuas just helping to send her to the bottom quicker.

The Norwegian king, crown prince, and cabinet boarded the Royal Navy Cruiser HMS *Devonshire* and departed on June 7 to form a government in exile. The Norwegian military held out until June 10, 1940, when they surrendered to the Germans. During Nazi occupation, the Norwegians lived under tyranny for five years. After the end of the war the population worked to put its lives back together, and Ju-88A-1 Werke no. 0880119, U4+TK, became forgotten, like so many other implements of war.

Aircraft Recovery

In the decade following the end of World War II, rumors began to spread about aircraft at the bottom of Jonsvatnet Lake. This body of fresh water is the drinking water source for Trondheim, thus explorations were quite limited over the years. In the 1980s, searchers were able to use side-scan sonar to map the bottom of the lake, and then deploy primitive ROVs to investigate the contacts. In 2003, the

Towed to shallow water, U4+TK is being stripped to lighten the load before lifting the bomber onto shore. Here the main landing gear has been removed and once on shore, the wings will next be separated from the fuselage. *Guttorm Fjeldstad*

Surviving Junkers Ju-88 Bombers

Model	Werke No.	Fuselage code	Location
Ju 88A-1	0880119	U4+TK	Norwegian Armed Forces Aircraft Collection, Gardermoen, Norway
Ju 88A-5	1379	HH+4V	Auto and Technical Museum, Sinsheim, Germany
Ju 88A-5	3386	4D+HM	Private Collector, United States
Ju 88R-1	360043	D5+EV	RAF Museum, Hendon
Ju 88D-1	430650	FE-1598	National Museum of the U.S. Air Force, Dayton, Ohio
Ju 88C-2	0881033	4D+HA	Norwegian Armed Forces Aircraft Collection, Gardermoen, Norway
Ju 88D-1	0881203	4N+EH	Norwegian Aviation Museum, Bodo, Norway
Ju 88A-4	0881478	4D+AM	Norwegian Aviation Museum, Bodo, Norway
Ju 88G-1	0880797	4D+HA	German Technical Museum, Berlin, Germany
Ju 88A-5	0886146	CF+VP	German Technical Museum, Berlin, Germany
Ju 88A-4	unknown		Private Collector, Belgium
Ju 88G-1	714628	2Z+BR	German Technical Museum, Berlin, Germany

Opposite: Ju-88A U4+TK is lifted out of Jonsvatnet Lake. The fuselage codes are clearly distinguishable. *Guttorm Fjeldstad*

underwater search firm Saastad A/S was brought in with state-of-the-art ROVs and camera equipment to investigate potential aircraft crash sites in the lake. It was during this survey that Ju-88A U4+TK was positively identified.

The recovery team had to deal with a large amount of paperwork and ensure there were no oil or fuel leaks into the water. In addition, they were given a limited amount of time to recover the aircraft. They were after two planes: the Ju-88A and a Heinkel He-111 bomber. The first aircraft recovered was Heinkel He-111H2 Werke no. 2320, fuselage code 6N+NH, which was recovered on September 2, 2004. The bomber last served with 1 Squadron, Bomber Wing 100 (1./KG 100), and the plane's last pilot, ninety-six-year old Artur von Casimir, was on hand to see the Heinkel come to the surface. Von Casimir told of the days when Josvatnet Lake was used as an airfield during the early stages of the Norwegian invasion. A few days after the fuselage was recovered, the Heinkel bomber's engines and propellers were brought to the surface, along with the tail section of Ju-88 Werke no. 113. Heinkel He-111H2 is now located at the Technical Museum in Berlin, Germany.

The recovery of Ju-88A U4+TK began on September 5, 2004. The aircraft was inverted, resting on the bottom of the lake. It was missing its tail and the magnesium engine mounts had decayed,

Status	Notes
R	
PV	Wings and tail used in replica display aircraft
S	Recovered by Jim Pearce
PV	Nightfighter variant
PV	Tropical configuration. Flown by a Romanian pilot into British hands. Evaluated by the USAAF.
PV	Displayed in as-found condition
R	
PV	Displayed in as-found condition
R	
R	Recovered from Norway. Early stages of restoration accomplished at Norwegian Armed Forces Aircraft Collection, Gardermoen, Norway.
S	
R	Recovered from lake in Hungary.

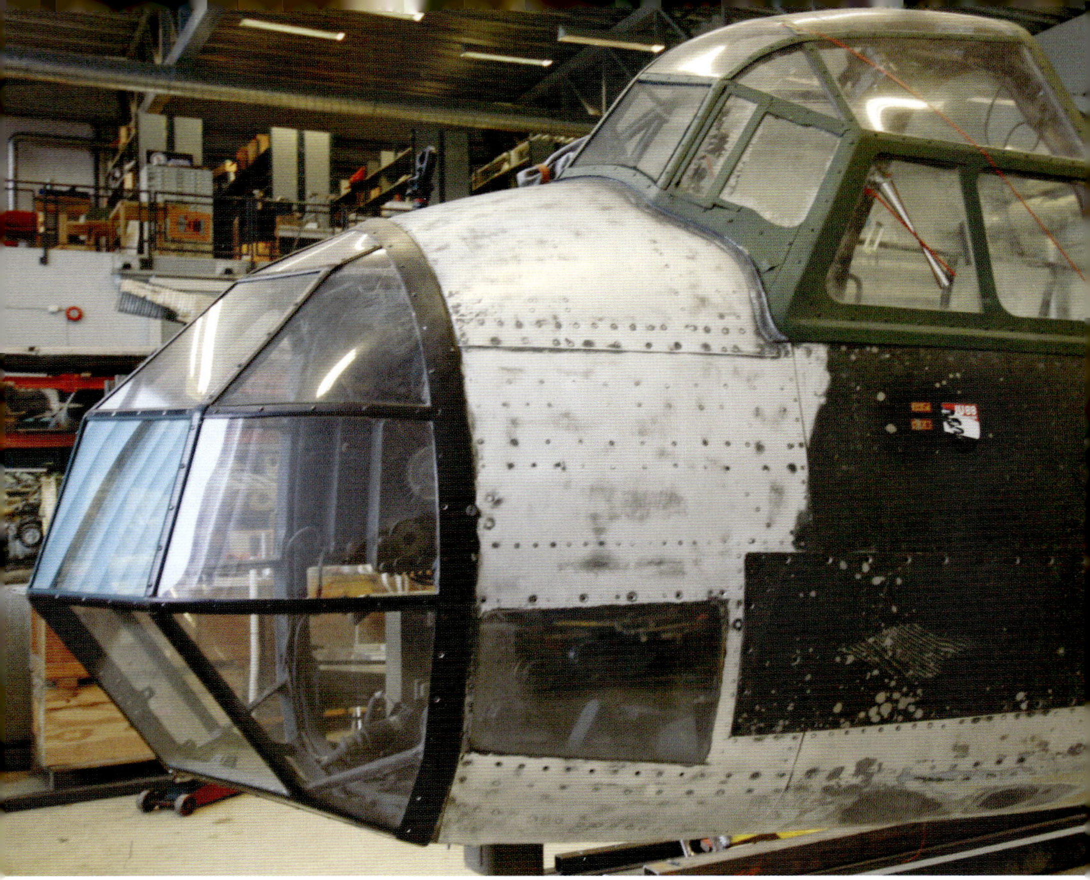

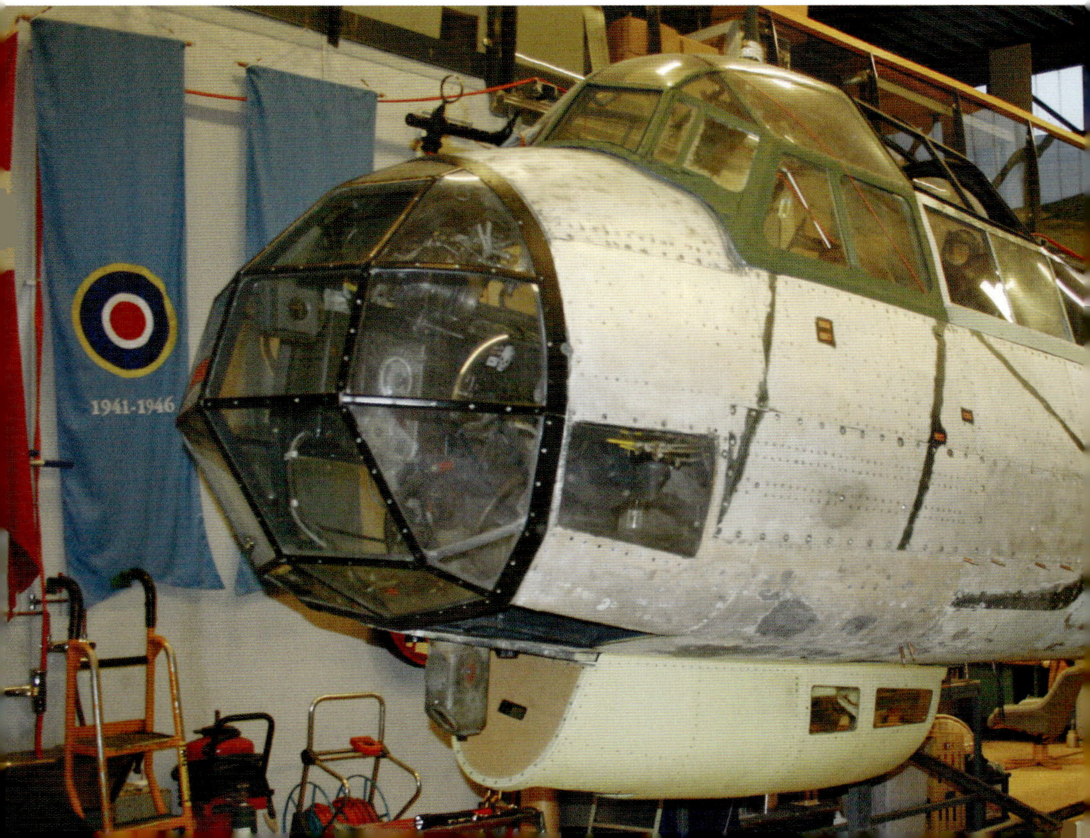
1941-1946

The canopy and nose section after glass has been installed and before and after the gunner's cupola has been fitted to the underside of the fuselage. Because the aircraft landed inverted when it sank, the gunner's cupola was not crushed. *Guttorm Fjeldstad*

dropping the powerplants to the bottom. On September 14 and 15, the engines were brought to the surface. The recovery team also investigated the crash of an Arado Ar-196, but left it alone when it was determined that the entire crew had perished in the crash and it was now considered a war grave.

After recovery from the lake, the Ju-88A hulk was transported to the Gardermoen Air Force Museum where it is now undergoing restoration to static display standards.

Restoration Progress

Nearly a decade into the restoration, a great amount of progress has been made. The most visible component is the Ju-88A's fuselage and cockpit. As the aircraft sank under the ice, nose-first, access to the cockpit and bomb aimer's position was blocked by the freezing water. The nose-down angle prevented the crew positions from being salvaged and souvenired, thus these stations were virtually complete when the aircraft was brought to the surface.

The tail landing gear and vertical tail and rudder, salvaged from a Ju-88C wreck, have been restored and installed into the fuselage as of the summer of 2013. At this time, the horizontal stabilizers are in jigs being rebuilt.

The restorers' craft is displayed in the Ju-88's cockpit where all of the original components have been meticulously overhauled and reinstalled. Notice the gunsight and in the lower center the aircraft's bombsight. *Guttorm Fjeldstad*

Because the aircraft flipped upside down during its descent to the bottom of the lake, the gunner's cupola was not crushed, and has now been restored and reattached to the underside of the fuselage near the cockpit. All of the cockpit instrumentation and details have been reinstalled, down to the Revi gunsight. As the pieces are brought together, the aircraft's original markings are applied.

Both 1,200-horsepower Junkers Jumo 211B twelve-cylinder, fuel-injected, in-line engines arrived at Gardermoen in excellent condition, and the all-steel propeller blades emerged from the fresh-water lake in relatively good shape as well. The exhaust stacks will need to be replaced and the restoration team hopes to have one of the powerplants running by the end of the rebuild.

The wings took the brunt of the strafing from the Skuas and a lot of the internal structure is heavily damaged. Wings from a wrecked

Ju-88C that was fitted with Ju-88A wings will be used as a source of structure for the rebuild. The bomb racks, dive brakes, and main landing gear have received the restoration team's attention and are awaiting installation when the wings are complete.

Although completion of the restoration of Ju-88A-1 Werke no. 0880119 is still a number of years in the future, once achieved, this former Luftwaffe medium bomber will join the museum's collection, which includes the sole surviving Northrop N-3PB, a Heinkel He-111, a Junkers Ju-52/3m, Spitfire PR XI, along with a number of ex-NATO fighter types ranging from the F-5 Freedom Fighter and Republic F-84 Thunderjet to a Lockheed TF-104 Starfighter.

The fuselage of Ju-88A U4+TK sitting in the sun on October 12, 2013. The vertical tail came from the aft section of a wrecked Ju-88C. The correct markings have been applied to the vertical tail. *Guttorm Fjeldstad*

Dyke Lake B-17 Recovery

Christmas Eve 1947 and B-17G 44-83790 rests on the ice-covered Dyke Lake in Labrador, Canada. Temperatures got down to -20 degrees Fahrenheit at night. *Chester Karney via Underwater Admiralty Sciences*

Sometimes things just don't go as you'd planned or hoped. That's been Don Brooks's experience when it comes to the Flying Fortress. He's had one crushed out from under him and another catch fire in flight, but the successful Georgia businessman is undeterred in his quest for a flying B-17. He wants one he can share with the public to commemorate the sacrifices of so many during World War II and as a tribute to his late father, Elton Brooks, a tail gunner on the 390th Bomb Group's B-17 *Liberty Belle*.

The B-17 that was crushed out from under him was B-17E 41-9101, *Big Stoop*, which Brooks helped to locate 265 feet under the Greenland icecap. His Liberty Foundation restored and operated the B-17G 44-85734 *Liberty Belle*, which suffered an in-flight fire in July 2011, and he was the driving force behind the recovery of B-17G 44-83790 that belly-landed onto the ice of frozen Dyke Lake in Labrador, Canada.

Formative Years

Having a father who participated in the fight against Hitler meant young Don Brooks had a lot of interest in World War II aviation. His father had always encouraged him to learn to fly, and had recommended that an aircraft would have been a great tool in their auto parts business. Elton Brooks passed away in 1978, and Don began working on his pilot's license later that year. "When I started flying, it was more because it was a good means of transportation for someone who lives in South Georgia," he explained. "Driving from where I live, you're three hours from anywhere. If you want to travel, flying's a great way to do it.

"The more I got into flying, the more interest I developed in World War II aviation and warbirds. I started going to some airshows, and I developed some friends that had warbirds. Somewhere along the line, it just kind of clicked that I'd like to try to find a B-17, restore it, and fly it around the country so other people could see it. At that time, there weren't as many flying as there are now."

Brooks was out of high school before he saw his first B-17. It was one of the fire ant sprayers operated by Dothan Aviation of Dothan, Alabama, which flew over Brooks's hometown spraying Mirex to combat the pests.

Search for the *Big Stoop*

Brooks decided he would find a B-17, learn to fly it, and barnstorm it around the country so people could see their aviation heritage in action. Brooks was flying a Bonanza for business use and he had become friends with Pat Epps of Epps Aviation at Peachtree-DeKalb Airport in Atlanta, Georgia. Brooks had read about Epps's trips to Greenland to look for the pair of B-17s and six P-38s that had landed on a glacier in 1942—the eight planes that became known as the "Lost Squadron" (see *Hidden Warbirds*, "P-38 *Glacier Girl*," page 124).

"I read an article in the Atlanta paper in fall 1988, describing how those planes had been found by a group of people out of Atlanta known as the Greenland Expedition Society. As I continued reading, I saw that Pat Epps had led the group. I knew Pat, and that was really interesting to me. I thought about getting involved and possibly obtaining one of the B-17s because there were not many on the market. I contacted Pat and participated in the next three expeditions (1989, 1990, and 1992). I was in charge of logistics, and I really enjoyed that," Brooks said.

They had found the airplanes using subsurface radar during their 1988 expedition, but they never actually saw a plane or touched it. In 1989, their primary goal was to produce a piece of one of the airplanes as tangible evidence that they had located the planes. That would gain them a three-year extension on their salvage rights from the government; at that time, the Danish were in charge (the land has since been ceded back to Greenland).

To obtain proof that the group had indeed located the aircraft under the ice, they developed a drill with a hole-saw that had a retainer clip on the bit. The team drilled into the plane and brought the metal sample to the surface. That three-inch diameter piece of aluminum skin was all the team needed to extend its salvage rights. On that 1989 expedition, Brooks was responsible for designing much of the equipment and shelters that were used.

The one thing the Greenland recovery team would need for following expeditions was a larger transport aircraft. Up to that time, they had been using single-engine and light-twin aircraft fitted with skis, but they'd need something larger to lift the P-38 components off the ice cap. "They were looking at buying a C-47/DC-3 that had a cargo door and skis," said Brooks. "They had located a plane that was in great condition and were going to buy it. Of course, I was a newcomer to the expedition, so I volunteered to buy the airplane and lease it back to the Greenland Expedition Society.

"As a side note, I had come up with an idea on how we might be able to eventually recover an airplane, get down to it. I designed and built the first meltdown machine, to make a shaft large enough that a man could go down in the hole. I designed that here in Douglas, Georgia. We took it with us on the 1989 expedition. Once we had accomplished our primary goal of getting the evidence that we had found the airplane, then we were free to try this machine. The principle worked, and we made notes on what we could do to make it better. When we got home, we all got together and compared notes. Then, we built a better version for the 1990 expedition."

In 1990 the group bored down to the B-17 *Big Stoop*, 265 feet below the surface. The team melted a cave around the nose of the bomber, then pumped out the water and went down to take a look.

Once safely down, the crew of B-17G 44-83790 and its passengers moved to the shoreline to build a camp and start a fire. After spending a cold night in the wilderness, the following afternoon, December 25, 1947, another search B-17 dropped emergency kits to the crew. They spent one more night before being rescued the following day. Pilot Chester Karney took this picture postcard view of 44-83790 sitting on the ice. *Chester Karney via Underwater Admiralty Sciences*

What they found was disheartening, especially for Brooks, because the B-17 had been crushed by the weight of the ice. The fuselage was only four feet high as the structure was not strong enough to withstand nearly fifty years of ice accumulation. Brooks brought back a number of B-17 parts—most as souvenirs, and many that are flying on B-17s today.

In 1992, the Greenland Expedition Society, with Don Brooks again on the team, returned to the ice cap and successfully recovered Lockheed P-38F 41-7630, today known as *Glacier Girl*, and owned and flown by Rod Lewis of San Antonio, Texas. Unfortunately for Brooks, his dreams of a recovering the B-17 *Big Stoop* were literally crushed out from under him.

The Dyke Lake B-17 and *Liberty Belle*

Before the 1992 expedition to the Greenland ice cap, Brooks began studying Air Force crash reports. Having learned a lesson with the search for *Big Stoop*, Brooks developed a set of criteria for the next

Twenty-five feet below the surface of Dyke Lake, one of the recovery team divers examines the cockpit. The B-17's unique throttle levers can be seen to the right. *Courtesy Underwater Admiralty Sciences*

B-17 recovery candidate. First, Brooks had to be able to drive to it in North America. Second, it could not be in salt water. And third, the crew had to have "walked away" from the crash, which would indicate that the plane would have been in pretty good condition after the crash.

Working with Georgia-based warbird restorer Tom Wilson, Brooks identified a recovery candidate that had belly-landed on Dyke Lake in Labrador, Canada. During the summer of 1992, while on a break from the Greenland Expedition Society trip, Brooks and Wilson traveled to Labrador and found a guide to take them by boat to the B-17's reported crash site. B-17G 44-83790 had belly-landed on the lake on December 24, 1947. The purpose of the trip was to see the conditions a team would have to work in. If they saw signs of the Flying Fortress, all the better.

"When the plane went down, that area was not settled at all," said Brooks. "Generally, nobody knew that the airplane was even in the lake. It's a big area now. There's some mining in the area, and that helped develop the region. The lakes in the area are also great for fishing. The main business up there is fish camps, where you fly in and they take you out on guided fishing trips to try to catch brown trout. The fishing guides have fished the entire area, and they didn't know there was an airplane down there. They thought we were crazy when we mentioned a four-engine airplane was down in the lake."

Brooks and Wilson decided it would be a good area in which to work, and a place that would be relatively easy to supply with tools, food, and crews. In addition, if the aircraft was recovered, it wasn't too far from the nearest road, and the parts could be trucked back to Brooks's facility in Douglas, Georgia, rather than having to airlift them off of an ice cap.

"We went to St. John's and talked to the provincial government. They were nice and said they were going to try to help us," Brooks said. "Of course, they kept putting us off through three administrations. We couldn't get permission to recover the airplane. They finally said, well, there was a law, a federal law that prohibited them from giving us the salvage rights, because if it had been there over a certain length of time, it was considered an historical asset, kind of like dinosaur bones. That protected them from being disturbed.

"In 1998, we were cleared to go up and see if we could find the airplane and take pictures of it, but we couldn't bring back any part of it. We couldn't disturb it. My thoughts were, 'Well, the only way I'm going to get this plane is to go through a legal procedure in the courts. If I'm going to spend that kind of money and time, I want to at least know the airplane is worth fighting for.' That's when I got Robert Mester and Mark Allen of Underwater Admiralty Sciences [UAS; Kirkland, Washington] involved. Both men are experts at underwater search and salvage. UAS planned the recovery expedition, and I actually dove down and took a look at the plane myself. We all agreed that the aircraft was worth saving. It was getting worse every year, but there was enough to make it worthwhile. When I returned home I contacted one of the top salvage attorneys in the world, David Horan, in Key West, Florida."

Surviving B-17 Flying Fortresses	
Operational	10
Static Display	23
Under restoration to operational	5
Under restoration to static	4
In Storage	6
Total Complete Airframes	48
Partial Airframes	7

Source: Scott Thompson, Aerovintage.com

Brooks had followed accounts of Mel Fisher's adventure to find the Spanish treasure ship *Atocha*. When Fisher found the treasure, everyone staked a claim to it. Fisher turned to David Horan to protect his find. Brooks had a meeting with the attorney and laid out the story of the Dyke Lake B-17. Horan was a former naval aviator and had a lot of interest in World War II history. He mentioned that there was a treaty between Canada, Great Britain, and the United States, which was still binding, that would allow for the recovery of the aircraft. Horan was on Brooks's team that afternoon.

Although the issue went to court, it wasn't a fight. It was more an attempt to discuss a point of law and to come to a resolution. The provincial government interpreted the law to read that anything older than a certain period was considered an antiquity, while the Canadian Federal government sided with Brooks and Horan that the treaty allowed for the salvage of this aircraft.

Horan won the case and phoned Brooks, suggesting he move immediately to recover the Flying Fortress. Thirty days later, the bomber was dismantled, on trucks, and heading for the Canadian/United States border.

A New *Liberty Belle*

The Dyke Lake aircraft was Brooks's second attempt to salvage a B-17, but it was going to be a long rebuild and at least a decade before the aircraft took to the skies. In addition, there was a time when it looked like Brooks would not prevail in the Canadian courts, and he began looking for another option to acquire a flyable B-17.

Tom Reilly of Kissimmee, Florida, was assisting on Brooks's Curtiss P-40 restoration. Reilly was restoring a B-17 for another customer who ended up backing out of the project. As Reilly searched for a new owner for the half-finished Flying Fortress, he approached Brooks. A deal was made and the Liberty Foundation was organized to acquire, manage, and operate B-17G 44-85734. When the restoration was complete, the aircraft was christened *Liberty Belle* in memory of the 390th Bomb Group B-17 that saw Elton Brooks safely through World War II.

"We flew *Liberty Belle* for the first time shortly after we'd returned from Dyke Lake. That gave me two B-17s, which was never my intent, but I did want to see the project through of rebuilding the Dyke Lake airplane," said Brooks. "We toured *Liberty Belle*, and it paid its own way. It enabled me to accomplish my goal of actually touring a B-17 around the country. We even flew *Liberty Belle* up into Newfoundland and Labrador to give the people up there a chance to see a B-17 up close because during the war, they had thousands that went over the area on their way to England and back. We toured Newfoundland and

Want to See or Fly in a B-17 Flying Fortress?

Many of the flyable B-17 Flying Fortresses tour the country each flying season.
Visit their websites for tour schedules.

Model	Serial	Name	Operator	Website
B-17G	44-83546	*Movie Memphis Belle*	The Liberty Foundation	www.libertyfoundation.org
B-17G	44-83575	*Nine-O-Nine*	The Collings Foundation	www.collingsfoundation.org
B-17G	44-83514	*Sentimental Journey*	Commemorative Air Force	www.azcaf.org
B-17G	44-85740	*Aluminum Overcast*	Experimental Aircraft Association	www.b17.org/tour
B-17G	44-85829	*Yankee Lady*	Yankee Air Force	www.yankeeairmuseum.org
B-17G	44-85718	*Thunderbird*	Lone Star Flight Museum	www.lsfm.org

Other airworthy B-17 Flying Fortresses:

Model	Serial	Name	Operator	Website
B-17G	44-8543	*Chuckie*	Tillamook Air Museum (Erickson Aircraft Collection)	www.tillamookair.com
B-17G	44-85784	*Sally B*	B-17 Preservation	www.sallyb.org.uk
B-17G	44-83563	*Fuddy Duddy*	Lyon Air Museum	www.lyonairmuseum.org
B-17G	44-83872	*Texas Raiders*	Commemorative Air Force	www.gulfcoastwing.org

Labrador because there were so many people in the province that had tried to help us get those salvage rights. Even the provincial government in St. John's were delighted to see the airplane. We left there on good terms. They're good friends and good people. We accomplished our goal.

"We flew *Liberty Belle* back to England in 2008. That was another one of my dreams, to go back and visit the base that the original *Liberty Belle* flew out of during World War II. We accomplished all those things and had a great trip. We brought the airplane back, and never had any problems with it. Eventually, we had the in-flight fire, and the firemen couldn't get to it to put the small fire out. We sat there and watched it burn.

"We've been searching the world for B-17 parts so that we can repair *Liberty Belle*. We intend on rebuilding *Liberty Belle* and flying it again."

Searching for the Dyke Lake Flying Fortress

The Dyke Lake airplane was built at the Douglas Aircraft Co. factory in Long Beach, California, and accepted by the USAAF on June 11, 1945. The bomber, serial number 44-83790, was immediately flown to storage at Lubbock, Texas, where she sat in the sun for more than a year. In September 1946, the bomber was prepared for flight and assigned to Spokane, Washington. In May 1947, the Flying Fortress was given duty on the U.S. East Coast at Westover Field outside of Springfield, Massachusetts. It was later transferred to the 1385th Base Unit at Bluie West One, at Narsarsuag, Greenland.

On December 24, 1947, pilot Lt. Chester Karney was flying the bomber from Harmon Field, Labrador, en route to the Bluie Eight airfield at Thule, Greenland. In addition to the copilot, Lt. Jack Bullington, the plane carried navigator Lt. E. Mann, and five passengers: Lt. James Veleth Jr., Cpl. Dale T. Lemon, Pfc. Clifford W. Thorne, and two civilians: Robert Tyrer and J. D. Cleghorn.

Having radio and navigational issues, the crew became lost. In radio contact with Goose Bay and running low on fuel, Karney belly-landed the bomber on Dyke Lake, 220 nautical miles from Goose Bay, Labrador. Everyone survived the landing and radio contact was reestablished with Goose Bay to initiate a rescue operation.

The crash report cited the pilot's and navigator's reliance on the radio compass for navigational information as a contributing factor in the incident. Apparently, radio compasses of the time became unreliable when an aircraft was homing on a station more than 130 nautical miles away.

When 44-83790 slid across the surface of the frozen lake, the bomber had 347.4 hours total time on the airframe. It was practically brand new.

After landing on the ice, the crew grabbed their survival gear and headed for firm ground to establish a campsite. Meanwhile, search aircraft were dispatched from Goose Bay covering twenty thousand square miles that afternoon in search of the downed B-17. Planes searched into the darkness while the survivors bedded down for an uncomfortably cold night. The following afternoon, at 3 p.m. local time, Christmas Day, the crew of the Dyke Lake B-17 radioed that they could see one of the B-17s searching for them. Supplies were dropped to help make their camp a little more tolerable during the coming night when temperatures were expected to drop to -20 degrees Fahrenheit.

While the food was being dropped, a ski-equipped C-47 was being flown from Greenland to Goose Bay where the skis would be installed and the aircraft test flown prior to making a rescue attempt. On December 26, at 2 p.m. local time, the ski- and jet-assisted take-off equipped C-47 departed Goose Bay for Dyke Lake. The C-47 successfully rescued the downed airmen and they were back at Goose Bay by nightfall.

Fast-forward fifty years and Don Brooks put together a team to travel to Dyke Lake to survey the bomber to establish whether the aircraft was worth recovering. Brooks tasked UAS with developing a plan to find and later recover the Dyke Lake B-17G.

After a number of planning meetings, Mester and Allen put together a team for the location and survey trip. On July 28, 1998, everyone met at the Pittsburgh Airport, Pennsylvania, and loaded their equipment into Brooks's C-47A 44-92606 (RAF serial KG395—this aircraft towed Horsa gliders during the June 6, 1944, D-Day landings; in Operation Market Garden on September 17, 1944; and on March 24, 1945, as part of Operation Varsity).

The C-47 carrying the nine-man expedition cleared customs at Quebec City, Canada, and then flew to Schefferville, Quebec, where the team's gear was transferred into Twin Otters for the thirty-five-mile flight to Dyke Lake. Everything the expedition needed—including an inflatable boat, side scan sonar, all the communications, diving gear, underwater video equipment, as well as food—was flown in to the

The B-17 was lifted off the bottom of the lake using a custom set of airbags developed by taking measurements from the Museum of Flight's B-17F *Boeing Bee*. In order to maintain the structural integrity of the aircraft, the majority of the outer wing panel nuts and bolts were replaced underwater and the engines and bomb bay doors were strapped in place. The B-17 raft was floated more than eighty miles to a location where a crane could lift the hulk out of the water, and where it could be disassembled and loaded on trucks for the trip to Douglas, Georgia. *Courtesy Don Brooks*

All smiles, the team stands on the B-17 raft having successfully lifted and floated the bomber back to civilization—with civilization being a relative term. From the condition of the propeller blades the engines were turning when the aircraft was set down on the ice, and more than six decades of sitting on the bottom of the lake saw the top of the fuselage crushed by varying levels of ice. *Courtesy Don Brooks*

Dyke Lake base camp. It took two Twin Otters two trips each to reposition the supplies and gear to the remote camp.

"During our whole trip on this first phase, we only saw one other group of people go by in a boat. The rest of the time, we saw no other humans," said Robert Mester. "There was one little incident where we landed and the Inuit guide was waiting at this little fish camp. One of the things we brought along was bottled water. He was pretty shocked by that. He couldn't understand why we would fly water all this distance when we're surrounded by water, the freshest and clearest in the world. He'd reach down and drink water where the aircraft finally was found. The whole area is pristine and in fact, we ended up, all of us, drinking the water and nobody ever got sick."

Everyone has different expectations of what will be found when searching for an aircraft crash site. Mester said, "We had all the information historically about the aircraft setting down on Dyke Lake in 1947, and the survival efforts of Lt. Karney and the crew, and heard all of the stories surrounding the incident. The location of this aircraft was so well documented that we thought we'd be able to go right to the site, drop down under the water, and see the aircraft sitting there. We assumed that the aircraft just melted through the ice and went to the bottom right where it crashed. Well, that was not the case."

Flying into the area, it was plainly visible to the team that from the water's edge to the beginning of the forest there were 150 feet of downed trees. This indicates that there is ice movement and that Dyke Lake is actually a wide spot in a connecting waterway. Somewhere downstream a dam had been installed which raised the water level of the lake. Depending upon the water's depth, the sonar equipment brought on the trip could have problems finding something deeper than 100 feet. (Sonar could work to 10,000 feet, but diving was limited to a maximum of 120 to 150 feet. At a depth of 120 to 150 feet, divers had only minutes of bottom time and needed to spend a great deal of time in lengthy in-water decompression holds. With no facilities or emergency transportation available at the Dyke Lake dive site, any problems in decompression could be fatal.)

Search operations got underway on July 29, 1998. While the inflatable boat was towing the side-scan sonar, other members of the team used aluminum boats rented from the nearby fish camp to explore the far banks of the lake. Here the team was able to locate the stranded airmen's camp. Their survival tents, food tins, and other equipment sat in the same place they had been abandoned in 1947.

"The conditions of the water out on the lake were such that you wouldn't be able to see much. It was very turbid. It was fairly flat until a wind picked up and it would get very, very rough and, of course, the mosquitoes were pretty much the size of P-51 Mustangs," Mester said. "It was pretty horrific trying to get out of the way of mosquitoes, the bugs, and go and do the work that you needed to do.

"The sonar found the communication cable that was trailing from the B-17 when it crashed. When Karney's crew figured out they needed to determine their location and to transmit their position to rescuers, they deployed this long cable out of the tail that works as an antenna to boost their signal. We found that antenna underwater more than one hundred feet down.

"That find reassured us that we had found the crash site and we expanded the search area after each and every day. It was getting a little disappointing. We were covering a lot of area. We were spending twelve to fourteen hours of daylight just expanding the search. We found one target that I suited up for and went down to take a look. We thought maybe it was one of the four engines and in fact, it was a large rock that was covered with a lot of marine growth. We searched a tremendously large area, expanding it the direction we felt the aircraft would've traveled based upon just our knowledge of the layout and the way that the water was moving. It was getting pretty depressing after more than a week.

"After a whole lot of searching, the days were going by without finding anything. Our budget was being consumed, and our allotted time at the site was growing shorter and shorter. I spent quite a bit of time with Don Brooks and the rest of the team looking at the charts and the information, and trying to make some determination of where we needed to look. We were running out of time. What are we going to do? We've got 'X' amount of time. How does Don Brooks want to use that time to look? One of the team had heard a comment about an old French fish camp that was some ways from the site and that there was talk a long time ago about a large American bomber near this fish camp site. In fact, somehow a seat from inside the aircraft had been recovered, and was over at the fish camp.

"It was decided two nights before we were running out of time that we would load up everything and float some twelve miles away from where we were and take a look at this fish camp site. Everybody was pretty tired, our food was running low, and they were just upset that we weren't having any luck at finding the bomber. We took off after breakfast and headed to the site."

To add even more doubt to the already discouraged team, they had to offload some of the crew and all of the equipment from the boats, and then send them down a set of rapids. Here the lake narrowed down into a shallow, rough waterway that was difficult to cross. Everyone wondered how an airplane as big as a B-17 could have possibly made it across the rapids without leaving chunks of metal behind. The team went past the fishing camp a little way to set up their new base of operations in the area where some said the bomber had come to rest.

"We set up the equipment and I explained to the team where I wanted them to search. It was more of just visually looking at the area and trying to determine: if an airplane came by here, where would we look?" said Mester. "Two teams were set up: one worked the sonar, and the other team searched the shoreline for any wreckage that may have ended up along the transition between the ice and the tree line. A couple of us then went over to explore the fish camp and we found the back bracing of a navigator's seat from a B-17, which Don recognized, just laying at an old fire pit near a dilapidated building.

"While the hours were going by, I watched the scanning crew and they started doing something that is out of the norm. You plan your scan and you scan your plan, which means you constantly move in expanding areas. They were coming back to one site, slowly. They were coming back at different angles and I knew they had something. While this was happening, Don was getting pretty discouraged. He spent a lot of money and things weren't going well, but everybody tried. It was a great effort. We were all alive and we all got along really well.

"Suddenly the boat broke from the search. The little Zodiac started heading back toward me. Our team has a little signal for when we have success finding something, and that signal was passed to me. I called Don down to the shoreline and the guys handed me the side-scan sonar records. I looked at it and it was just like a Polaroid picture of a four-engine bomber with its nose, wings, and aft fuselage. I motioned to Don saying, 'We've found something here, but the guys think it's a log.' I handed the sonar records up to him while the crew was watching Don's expression. Don looked at the records, his eyes got gigantic, and he said, 'That looks like our B-17!' We had, in fact, found a B-17 there.

"We weren't sure it was the Dyke Lake B-17 because other B-17s had gone down in this area as well, a number of them, but we were all pretty pumped up. We planned to go back the next day and make a dive on the site to determine if it was '790.

"The next day we made it through the rapids down to where the B-17 was located, threw a marker out, and went to shore. I suited up and headed out with one of the other divers who had a still camera and I had a very advanced underwater video system. The current was so bad when we entered the water that my other diver shot through the area and missed the plane. I was fighting the current and all of a sudden I'm at the front end of the cockpit near the left wing of a B-17, and it was an amazing sight. It was loaded with all types of fish and it was clean. The paint was still on the wings and the top turret had been removed on this aircraft, which gave us access to get inside the cockpit. I wasn't able to see a number because I was back near where that tail section should have been. At some point it had been torn from the wreck.

"I managed to squeeze in through the top turret opening, which put me right behind the pilot and co-pilot seats at the entrance to the bomb bay. I was videotaping through the windshield and shooting all of the gauges and controls, and I got the strangest feeling. It was one

After disassembling the bomber, it was loaded onto trucks for the long, cross-country drive from north of the Canadian border to Douglas, Georgia. The U.S. national insignia looks to be in excellent condition considering the time spent under water. *Courtesy Underwater Admiralty Sciences*

of those eerie feelings like somebody's watching you; a disconcerting feeling. I turned around and I looked in the bomb bay and here is a coffin that's open and I knew then that we'd found '790 because in the crash report, they listed that they were carrying two deceased individuals as cargo; one was a doctor, another was an airman. The doctor was hunting walrus in Greenland and fell through the ice and drowned. The other was a military airman that was hanging his wet socks on some pipes or wires near his bunk and electrocuted himself. They were being transported in the airplane and when the rescue came, they actually removed the bodies and took them out with the people that they rescued.

"Now that the aircraft had been located, my purpose was to check the condition of the aircraft, especially at the wing root, to determine if it was a candidate for recovery and a candidate for recovery to restoration."

Don Brooks had taken a dive course and had brought a dry suit on the trip. He suited up and pulled himself out to the wreck using the divers' excursion line that stretched from the shore to the aircraft. Brooks got to see the aircraft he'd been searching for up-close, under water.

Mester and Allen determined that at the time of the crash landing, the lake's water level was higher. When the spring thaw melted the surface ice, the B-17's fuselage and wings reflected the sun, so that eventually it was floating on an ice raft approximately three feet thick and the same shape as the bomber's shadow. It began to float downriver, moving with the circulation of the lake. Sitting on its own raft, the B-17 shot the rapids, if they were there at the time, or just floated down to the next wide spot where it settled and eventually sank to the bottom.

Recovering the B-17

Having established that the aircraft was worth recovering, it took four years for the ownership issue to be settled. By 2004, the permits were in hand and Mester and Allen immediately got to work. The B-17 was thirty-five miles from Schefferville by air, but nearly a hundred miles from the nearest major road—if you could call it a road. The plan was to tow the aircraft nearly ninety miles downstream to an area where they could dismantle the aircraft enough to have a crane

lift the bomber onto the shore. From there, the bomber's components could be further reduced to fit into and on trucks for the trip to Douglas, Georgia.

"Each of the phases of this project—first of all going to find it and then actually locating it, determining if it's a good one, the legal process, securing the rights, designing and developing a recovery plan, and then actually executing that recovery plan—are a major undertaking. You just try to prepare yourself and the crew for anything that could possibly happen," Mester said.

In designing the recovery, the team faced quite a few unknowns, such as water depths and currents along the route. While all of the World War II aircraft were maintained using forklifts and other heavy equipment, at the recovery site the B-17 would have to be moved and dismantled using lift bags and hand tools. There would be no power tools available under water at Dyke Lake.

"One of the first things we did was to get permission from the Museum of Flight in Seattle to examine its B-17 *Boeing Bee*. We were able to remove a number of fairings to determine the best way to disassemble and transport the Dyke Lake B-17," said Mester. "The problem we had to solve was how to pull the wing root taper pins that hold the wings to the fuselage. The folks at Seattle were a great help. Then we flew down to see Dave Goss at Gosshawk Aviation in Arizona, who was replacing the wing pins on a B-17 in his shop. We spent time there helping and observing and trying to figure out how we were going to do this work in the field.

"The aluminum did quite well, but all the other metals went through electrolysis and started sacrificing. We found out that when we got ready to start putting lift on the aircraft that the exhaust, the engine, engine mounts, the bomb bays—all the important things—would fall off the airplane. We actually had to disassemble the airplane and reassemble it underwater to stabilize it for the lift in order to transport it."

"When the provincial government finally agreed that we could go ahead and recover the aircraft, they gave us a short window to put a team together, to make all the arrangements, to go up there, and then do the recovery," said Mark Allen. "We were given from the middle of August to the middle of September to remove the bomber. That was a problem because Bob and I were going to do a reconnaissance of Dyke Lake to ensure that the aircraft was still where we found it, and to make some additional checks to verify that it was worth recovering.

We contacted our guide, Gary Shaw Sr., and he said, 'Don't bother coming up. The river's still frozen over.' We never got a chance to do the reconnaissance dive.

"We went ahead and finished prepping the team that was going to go with us, and then working and training everybody, getting the specialty lift bags made locally by Carter Lift Bags, in Enumclaw, Washington, to doing the training with the Museum of Flight. We had at least four sessions on that aircraft and they allowed us to go ahead and take fairings off and to take pieces off so we could actually see how everything looked. We had our team training on the museum's B-17 so we could identify all the access panels almost blindfolded because when we'd be working underneath the aircraft, it would be in the shadow and even with our guide lights, it would still be very dark. This knowledge would enable us to find and open access panels underwater in near darkness so we could reach up and find the main spar or the after spar to tie in the lift bags at a place where we're pulling from a structural component.

"Our first dive was late in the afternoon of August 17, 2004. We got up early that morning in Labrador City, piled all the gear in vans, and drove three and one-half hours to get to the dock where we met a high-speed boat that was going to take us on a two-hour ride up-river to our base camp. All our gear had already been transported to the camp and we arrived around 3:30 or 4 p.m. We found our dive gear, got suited up, and Bob dropped us right in on the plane. The bomber sat twenty-five feet below the surface.

"I know the B-17 has a 104-foot wingspan and is about 76 feet long. However, every time we went back to the *Boeing Bee* at the Museum of Flight to look at it and work with it, that plane just seemed to get bigger and bigger. When we dropped down on the aircraft in the lake, it seemed to be the size of a 747, and I was getting worried that we didn't bring enough lift bags. We had 70,000 pounds of lifting capacity in the bags we brought and the aircraft's dry weight is 39,000 pounds. We knew we were missing the empennage, so we figured we were going to have enough lift, but that plane sure looked big."

While diving on the B-17, the team began to assess a number of areas where they had planned to tie in lifting bags. One of the first places they inspected was the engine mounts. Securing lift bags to the tops of the mounts would give them greater control when lifting the plane off the bottom of the lake. But it only took one or two taps from a hammer to see the engine mounts crumble. The dive team also

The dive team drove out of town ahead of the trucks to catch a glimpse of the bomber going down the road. They said the clouds opened and the sun shined brightly as the B-17 headed down the road to a new future. *Courtesy Underwater Admiralty Sciences*

brought up a number of exhaust parts, and with input from what the team had seen, it was determined that the aluminum was in excellent condition, but all of the steel in the aircraft had corroded away. This was also a concern when the aircraft was lifted off the bottom of the lake as the team wondered if the bomb bay doors would drop back to the bottom because the metal hinge pins might be gone. Diver Zack Jones had to strip off his tanks and regulator, and, using a pony bottle and regulator, crawl underneath the aircraft carrying straps, which were then secured to ensure the bomb bay doors stayed with the plane.

Two weeks later, when they were able to lift the fuselage clear of the water, the team found out that had they not strapped the bomb bay doors they would have dropped to the bottom because the hinge pins were, indeed, corroded away.

When the dissimilar metals corroded away, it created a whole new set of challenges. For example, the bolts holding the outer wing panels were gone, and divers had to install more than eighty nuts and bolts under water to keep the panels in place for the long tow to shore.

Dyke Lake's water temperature in the middle of August was 55 degrees Fahrenheit, and the crew of twelve divers typically worked one-hour shifts before they were chilled to the bone. The operation began on the evening of August 17 and by the afternoon of August 23, the lift bags were rigged and ready to test. The team put in 118 hours submerged working on getting the aircraft prepared for lifting and getting it off the bottom.

"Originally, we were going to lift from above, but since the engine mounts were corroded and were not going to allow us to do that, we had to go to plan B. We had to try and figure out how we were going to get the lift bags underneath the nacelles," said Allen. "We were going to cradle the forward part of the aircraft inside the engines underneath the nacells and lift from underneath, but the airplane and the engines themselves were hard down on the river bottom, which was rock gravel.

"One of our divers, John Schliemann, said he could excavate underneath each engine to position the lift bags. These lift bags are pretty darn big. They're over eight feet in length and three feet in thickness when inflated. They were designed specifically for this project by taking measurements of the Seattle B-17 and having these bags custom-made to our specifications. I didn't believe it could be done. In two dives he was able to remove a substantial amount of material from underneath the engine. The divers then began excavating the area to get the lift bags underneath the nacelles and rigged so we could make a lift from underneath."

To provide lift to the fuselage waist area, divers rigged a girdle around that section to provide lifting points for the air bags. Things were pretty well set by August 22, and the final tweaks were made the next day. An anchor pattern was set to ensure the floating B-17 didn't get swept downriver in the current.

In preparation for the lift, the one big guide boat, the *Silver Dolphin*, was anchored off the bomber's nose. Slowly adding air to the bags, nothing happened for the first thirty minutes. Then suddenly the bomber lifted from the bottom and the nose broke the surface at a forty-five degree angle, wedging the aft end of the fuselage under a rock.

"We watched it pop out of the water. I was terrified as it came out because that's not what we had wanted, but that's how it came out. I was afraid something was going to break, or the wings were going to collapse," said Allen. "Any number of things could have happened, but the plane held fast and secure for us. That evening we went in to celebrate the start of getting the aircraft out of the water. When we woke up the next morning, August 24, and came out to look at the

The forward fuselage of 44-83790 comes together in the restoration jig at Don Brooks's Douglas, Georgia, restoration facility. Tom Wilson of the Hawk Factory in Griffin, Georgia, is building a new cockpit canopy and turtle deck for the Dyke Lake B-17. *Chuck Geise*

bomber, it had sunk under water again. Apparently it had turned with the current and the air leaked out of the lift bags. Our first question was did we shred any of the bags, and then how sound were they. We spent the rest of the day re-rigging the lift bags and getting ready to inflate them the following day."

It took two more days of lifting, having the load come up at some odd angle, dropping it back to the bottom, re-rigging, and re-lifting it before the airplane found an equilibrium. Once stable, the team began to pump water out of the fuel tanks, starting with the Tokyo tanks in the outer wing panels. The next task was to use the pressure washer to blast out as much river silt as possible from inside the wings and fuselage. As the tanks were pumped dry and the silt was removed, the aircraft and its raft began to ride higher in the water, drawing about five feet.

Once the aircraft was floating and stabilized, a night watch was maintained. If the bags began to deflate, the crew could be called to action to keep the plane from sinking again. Glow sticks were placed on the wing tips to help those on shore judge the stability of the raft. If the reflection of glow stick and the stick itself merged, there was a problem. Don Brooks and Gary Shaw Sr. stayed up the first three nights on watch.

When not watching the raft, everyone kept an eye out on the weather. They needed minimal winds for crossing the Smallwood Reservoir because that body of water could develop swells and waves that could potentially swamp the floating B-17. The team chose to transit the reservoir on August 29. The *Silver Dolphin* would lead, with guide boats maintaining a straight track, attached to the raft by ropes. There were numerous areas where the river zigged and zagged, and the flotilla could not enter water shallower than five feet while towing the bomber. It got pretty touchy in some areas because the depth of the river was only six feet.

"We got pretty good at towing at that point in time. We were developing some teamwork and getting a better handle on it. We had a couple of breakdowns and had a couple of hairpin turns to deal with," said Allen. "We kept towing and at some point we managed to get some pizza delivered to us from one of the fish camps. That tasted pretty darn good!

"We towed on into the night and through midnight. We had the system down really well and we were getting close to an area that had tight narrows to go through. The crew was doing so well in navigating the boat that we said, 'Alright, we're going to shoot this at night.' About that time, our radio communications began to fade because of batteries. We had no choice but to anchor up again in the middle of the river, wait for batteries to charge, and wait for daylight. It was another cold night, but not quite as cold as the night before.

"The next day we were about to enter Smallwood Reservoir and we noticed that the weather report and actual observations were showing that the waves were building. We were looking at swells two to three feet in height and the wind was building. We had a conversation between Gary Shaw, Don Brooks, Rick Snow—who was our boat coxswain—and

Don Brooks has Chuck Geise rebuilding both the Dyke Lake B-17 (44-83790, foreground), and the ex-20th Century Fox studios B-17G (44-83387, rear) that was used for the interior scenes of the 1949 film *12 O'Clock High* and the mid-1960s TV series of the same name. *Chuck Geise*

myself, to answer the question as to whether we ran and found a place to anchor until the weather gets better, or did we go for it?

"The consensus was that this maybe is good as it's ever going to get for the next two to three weeks if not for the rest of the year, so we had to go ahead and push on. We made it out and started getting across Smallwood Reservoir. It seemed to be going fine, even though waves were building and we had white caps out there. Then Hamilton Halford came up on the radio and said, 'Mark, you need to know that number three and four engines are beginning to move back and forth.' We were concerned about the integrity of the engine mounts and hearing that the two engines were beginning to move back and forth was extremely concerning because if we lost those engines they would take the lift system with them and we would probably lose the plane in 120 feet of water."

The decision was made to change course and reduce speed, and have the men riding on the bomber secure the engines as best as possible. Two men were on the aircraft at all times during the tow to maintain proper air pressure in the lift bags and to monitor the rigging during the transit. Having secured the engines and now traveling on an altered course at a slower speed, the tow continued until it reached its final destination, a point of land called Lobstick where the raft was secured for the night.

Most of the crew had been awake for more than seventy-two hours before they were able to get any quality sleep. The tow proceeded at one and one-half to two knots per hour, and it took fifty-eight hours of stressful towing to get the bomber to Lobstick where it would be disassembled and readied for shipment.

"The weather began to deteriorate after we landed," Allen said. "On September 1, the waves were six to eight feet on Dyke Lake. We were hoping to send a guy back up to the camp to bring the rest of our gear down, and we were hoping to go back and continue looking for the empennage while Don Brooks's team started to disassemble the aircraft. We weren't able to go back up and get our gear until more than a week later because the weather was so bad. As it turned out, the day we crossed Smallwood Reservoir was the last calm day for a couple of weeks. Had we stopped we would have missed our weather window."

Once the aircraft was on solid ground, Mester and Allen oversaw the construction of transport cradles for the aircraft, made arrangements to clear U.S. Customs at the border, then get the wide load permits through a dozen states, and the bomber safely delivered to

Douglas, Georgia. They say the paperwork is never done, and in this case they're right.

"Finally, on September 29, 2004— it was a Wednesday—we had the fuselage, the wings, the engines, and all the things that fell off the airplane that we had to salvage and recover in addition to the aircraft itself," said Mester. "All of these were finally loaded and stabilized on trucks and headed south. When the trucks headed out, we went out of town and climbed up on a hill on the road and were able to get a great view of the trucks coming with the aircraft on it. There were two of them. The sky opened up. The sun came out. The pictures just turned out beautiful. It was like the B-17 was wanting to make sure we had the right light to see the plane leave for its new home."

The Final Count

"When the in-flight fire happened on the *Liberty Belle*, we'd already started restoring the Dyke Lake airplane," Brooks said. "That's my priority right now: to work on the Dyke Lake airplane. In the meantime, I needed a lot of parts, so in 2013 we worked a deal with Bruce Orriss and bought that project and a lot of spare parts that will help us get one of our planes flying more quickly."

Need a score card?

Liberty Belle is B-17G 44-85734, and the Dyke Lake B-17G is 44-83790. The ex-Orriss B-17 project is B-17G 44-83887. If you're reading this book, chances are you've seen 44-83887 many times: it was used by Fox Studios to film the aircraft cockpit and interior shots for the 1949 film *Twelve O'Clock High*. The TV series of the same name in the mid-1960s used it, as did the filmmakers who shot *The 1,000 Plane Raid*. It then hung in a bar in Greeley, Colorado, for many years before it was acquired by veteran aviation propman Bruce Orriss who began the task of putting the bomber back together.

"I truly don't know right now what my ultimate goals are for the Orriss airplane, and I don't know what course I'm going to take with the Dyke Lake airplane," Brooks said. "I could build the Dyke Lake airplane to fly under its own number, or I could use parts from the Dyke Lake airplane to repair the *Liberty Belle*. As far as the Orriss airplane, it could easily be restored as a museum static aircraft. Or, if I want to take more time, put more money into it, it could be restored to fly. All I can say for sure is that I plan to have one flying B-17. Not sure which one it'll be. I might have as many as three."

Decisions. Decisions.

Lend-Lease P-39 Restoration

During World War II, the United States lend-leased the Russians 4,924 Bell P-39 Airacobras. The 25th Ferrying Squadron, 7th Ferrying Group of the Air Transport Command was responsible for delivering lend-lease aircraft to Nome, Alaska, where the Russian pilots would then fly them to the battlefields in Europe.

However, 178 of the P-39s that left the factory to serve with the Soviet air force never reached Alaska. These lost aircraft are a wreck-chaser's dream come true—somewhere between Buffalo, New York, and Nome, Alaska, lie the remains of these fighters.

Two Russia-bound Airacobras took off from Fort Nelson, British Columbia, Canada, on the morning of December 6, 1943. Second Lieutenant Hal Coons and his wingman, 2nd Lt. Delos R. Carpenter, were airborne at 11:05 a.m. MWT. Their first stop was Whitehorse, Yukon Territory, Canada, with a final destination of Ladd Field in Nome.

For this flight, Lt. Carpenter was flying P-39Q-15, serial number 44-2485. He had logged 475 hours flight time while in the Army Air Forces, 40 of those hours in the P-39. Once airborne, both planes climbed through the "high overcast" and headed toward Fort Nelson on course. One hour later, Lt. Coons decided to let down through a hole in the overcast. As the aircraft descended, the two became separated.

After making a number of 360-degree turns in an attempt to locate Coons, Lt. Carpenter, who was now low on fuel, decided to belly-land on a frozen lake. He jettisoned his drop tanks and landed the aircraft with only minor damage to the propeller, but the right wing had buckled. The aircraft broke through the ice; however, it remained afloat.

Lieutenant Carpenter removed his survival gear, which consisted of a shovel, an axe, rope, and smoke flares, and then headed to the shore, where he managed to build a fire. At approximately 8:30 p.m., he made contact with Lake Army Airways.

The following day at about 3:00 p.m., Lt. Carpenter was rescued by a ski plane. The USAAF reported that the ice was so thin and the lake so short, fuel had to be drained from the ski plane to get him out.

Lieutenant Coons had decided to set down because he had suffered a compass failure. Unable to locate Lt. Carpenter, he had followed a valley until he flew out from under the clouds, where he recognized his position as south of Fort Laird. Coons returned safely to his point of origin, Fort Nelson, to attempt the flight to Whitehorse another day.

Lend-lease aircraft built in the United States were collected by Soviet pilots at Ladd Field, Alaska. This July 9, 1943, shot of Ladd Field from the top of a hangar shows P-39s in the foreground, B-25s and C-47s in the upper center, and A-20 medium bombers in the upper right. Lieutenant Delos R. Carpenter was ferrying P-39Q 44-2485 to Ladd Field, but never made it. *U.S. Army Air Forces*

Finding and Recovering an Airacobra

Forty-seven years later, warbird restorer Gary Larkins of Auburn, California, set out in search of Lt. Carpenter's P-39. Larkins began by looking for the aircraft at the coordinates listed on the Air Force crash report. At the time of the crash, this area was uncharted.

The Air Force listed the crash location as 120 miles due north of the Smith River, so Larkins's first two trips were spent searching all the lakes north of the reported crash site. Although these first two

Lifting the fuselage, center section, and wings back into the air after more than forty-seven years. *Gary Larkins*

trips were unsuccessful, he secured the necessary permits from the Canadian Ministry of Transport—for the P-39Q-15 and two other P-39s—in anticipation of a successful recovery of this aircraft.

On Larkins's third trip he located 44-2485 in Carpenter Lake, 240 miles from Watson Lake, Yukon Territory, Canada. The aircraft was recovered on July 19 and 20, 1990, at the tail end of a ten-day trip into the Canadian wilderness.

Upon arriving at Carpenter Lake, Larkins and his crew immediately knew that they had found the crash site. The area was littered with fifty-five-gallon drums, the remnants of tents, wood stoves stenciled with U.S. Army Corps of Engineers, and a medical kit.

George Carter and Wayne Lloyd, both former U.S. Navy SEALs, worked with Larkins on the recovery effort. They are also the same divers who had dived through two inches of ice to locate a PBM Mariner in Chesapeake Bay, only to find that the aircraft was not worth recovering.

Larkins, Carter, and Lloyd arrived at the lake in a de Havilland Beaver fitted with floats. Aside from the regular camping gear, they brought a rowboat, six dive tanks, a pair of lift bags, tools, and a sonar scanner. Because of space limitations, they could not get an outboard engine on board the Beaver, but a set of oars fit nicely.

Their first two days at the site were spent rowing the scanner the length of the lake. Once contact was made with the aircraft, they began diving on the Airacobra. They placed a buoy on the fuselage and then began to search for other parts underwater.

Former U.S. Navy SEAL divers George Carter, left, and Wayne Lloyd, near the Airacobra's tail section after recovery. Carter and Lloyd unbolted the tail section under water to lighten the load when the helicopter lifted the wings and center section out of the water. *Gary Larkins*

Bell P-39Q Airacobra, serial number 44-2485, sits on a trailer at Watson Lake. The helicopter flight from the crash site to Watson Lake took more than nine hours and covered 240 miles in the air. The wings were covered with netting to spoil the lift as the fuselage and wings were flown from to Watson Lake. *Gary Larkins*

The nose of the aircraft was at a depth of sixty feet, and the tail was at thirty feet. While the aircraft was underwater, the divers removed the empennage. Carter and Lloyd had to work fast as they only had six air tanks, two of which were expended in the removal and recovery of the tail. Larkins raised the tail to the surface with a rope and then towed it to shore, stroke-by-stroke, with the rowboat. On shore, the horizontal stabilizer was removed and readied for transport.

Next they raised the airframe with lift bags. Again using only the rowboat, Larkins towed it into shallow water where it was rigged with a harness so that the helicopter could pick it up. Once rigged, the air was let out of the bags, setting the P-39 back on the lake bottom.

After finding the P-39, Larkins radioed for the helicopter. It took three days for the Bell 205 to reach the wreck site at the right time of day. Once over the lake, a diver in the water coordinated with the helicopter pilot to pick up the P-39. It took an hour to safely get the hook to the diver and onto the airplane.

After this hour of maneuvering, fuel for the helicopter was getting critical. Once the hook was secured to the P-39, the fuselage and wings were lifted onto shore. While buckets full of mud were removed and the harness was double-checked, the helicopter pilot took a well-deserved rest.

When the USAAF salvaged the aircraft, they had removed the engine, cockpit instrumentation, and the landing gear, flying those items out to Fort Nelson. In the process of salvaging the P-39, the USAAF had removed numerous pieces, setting them on the ice near the airplane.

When spring came, the ice thawed, and all the parts left behind sank straight down to a depth of sixty feet. Divers Carter and Lloyd recovered a number of inspection panels and the cowling, located the bent prop, and brought a number of other small parts to the surface.

The staff at Pacific Fighters went through the Airacobra with an eye to upgrading its external appearance. New canopy glass, a new air scoop, and many other areas were detailed. *Mike Oliver*

P-39Q 44-2485 underwent a second restoration in 2013 at Pacific Fighters in Idaho. The Lend-Lease Airacobra will be one of the aircraft on display at the new Erickson Aircraft Collection in Madras, Oregon, when that facility opens in late 2015 or early 2016. *Mike Oliver*

Netting was rigged over the fuselage and wings to foil the lift, and a drag chute was deployed for directional stability during the airlift to Watson Lake. Once the P-39 was ready to be transported, the Beaver was dispatched to drop fuel at two locations while en route to pick up the divers, their gear, and the P-39's empennage.

Transporting the P-39 over the 240 miles to Watson Lake took nine and a half hours. The Bell 205 was only able to make sixty knots and, at times, had to clear mountain ridges as high as 14,000 feet.

Once at Watson Lake, the wing was demated from the fuselage, and all the aircraft components were packed into shipping containers.

P-39Q 44-2485 is in excellent shape. Both the Bell dataplate and the Russian export dataplate are still in the aircraft. The Russian red star is still visible on the empennage and the wings.

The aircraft was taken to Gary Larkins's Auburn, California, shops for a restoration to stabilize the aircraft. Here an Allison engine was fitted to the airframe and the plane was transported for display at the Tillamook Air Museum in Tillamook, Oregon.

In June 2012, the aircraft traveled north and east to Pacific Fighters in Idaho Falls, Idaho. The Tillamook Air Museum collection is moving inland to Madras, Oregon, where a 65,000-square foot hangar is being built. The relocation should be complete by January 2016 when the museum's current lease expires. The Tillamook Air Museum is owned by the Erickson Group Ltd., which also owns the Erickson Aero Tanker operation, also at Madras.

Currently 44-2485 has been shipped from Pacific Fighters and is in storage awaiting display space at the new Madras facility. When the collection gets to central Oregon, it will be renamed the Erickson Aircraft Collection after its founder, Jack Erickson, who started the collection in 1983 and has acquired more than thirty-five rare and historic aircraft.

Check Out a P-39: Museum-Displayed Bell Airacobras

Model	Serial	Location
P-39D	41-6951	Beck's Military Collection, Mareeba, Queensland, Australia
P-39N	42-8740	Yanks Air Museum, Chino, California
P-39N	42-18814	Pima Air & Space Museum, Tucson, Arizona
P-39N	42-19027	Planes of Fame Air Museum, Chino, California
P-39N	42-19158	Novosibirsk Restorers, Russia
P-39Q	42-19597	Commemorative Air Force, San Marcos, Texas
P-39Q	42-19993	Lews Air Legends, San Antonio, Texas
P-39Q	42-19995	Buffalo and Erie County Naval and Serviceman's Park, Buffalo, New York
P-39Q	42-20000	March Field Museum, Riverside, California
P-39Q	42-20007	Virginia Air & Space Center, Hampton, Virginia
P-39Q	42-20442	Norwegian Aviation Museum, Bobo, Norway
P-39Q	44-2433	National Air & Space Museum, Chantilly, Virginia
P-39Q	44-2485	Tillamook Air Museum, Tillamook, Oregon
P-39Q	44-2664	Aviation Museum of Central Finland, Tikkakoski, Finland
P-39Q	44-2911	Niagara Aerospace Museum, Buffalo, New York
P-39Q	44-3887	National Museum of the U.S. Air Force, Dayton, Ohio
P-39Q	44-3908	Kalamazoo Air Zoo, Kalamazoo, Michigan
P-400	AP335	Military Aviation Museum, Virginia Beach, Virginia

Part Two

In from the Swamps and Jungles

There is a huge number of warbirds hidden in the jungles and swamps across the former battlefields of World War II. The swamps and bogs of the western Soviet Union have yielded all sorts of aircraft, both Allied and Axis. A Hawker Hurricane with its well-preserved pilot were recently recovered from a swamp in northern Russia, and, although the aircraft was heavily damaged, there was certainly enough of the fuselage and wing center section to rebuild the remains into a flyer. An Il-2 Sturmovik was recently recovered from a swampy area in the Pskov region as well.

In the Pacific, there are numerous wrecks in various states of preservation, many in situ, over-grown, and simply forgotten. In the 1970s, Canadian Robert Diemert had recovered a Val and three Zeros from the island of Ballalae, in the Solomon Islands. The Val is now under re-restoration at the Planes of Fame Air Museum in Chino, California, and two of the Zeros are now on display at the National Naval Aviation Museum in Pensacola, Florida, and the Pacific Aviation Museum at Pearl Harbor, Hawaii.

In recent years, there was another attempted recovery from Ballalae, which had an aircraft graveyard of sorts at the center of the island. This recovery was done at the behest of the Solomon Islands National Museum, and substantial pieces of a Val, Betty, Oscar, and a couple of Zeros were collected, but an ownership dispute stopped their removal. In addition, a number of American naval aircraft, including a couple of SBDs, TBFs, and a Corsair, had crashed on the island. This small island holds a large number of wrecks and is just across the channel from Guadalcanal.

From New Guinea, Fred Hagen and Robert Greinert recovered a Republic P-47D Thunderbolt, serial number 42-22687. The fighter

Yap Island holds a number of World War II wrecks including this Mitsubishi G4M1 Model 11. This tail section rests twenty yards away from the main fuselage. Craig Fuller, the founder of Aviation Archaeological Investigation and Research, was part of a June 2006 TIGHAR the International Group for Historic Aircraft Recovery) survey of four World War II Japanese aircraft wrecks on the island. The U.S. National Park Service provided a grant to the Yap Historic Preservation Office for the survey to determine a management plan for the aircraft wrecks. *Craig Fuller*

was being flown by 1st Lt. Marion C. Lutes, who crashed into the jungle on the side of an 8,200-foot mountain near the village of Nando. In October 2004, the wreckage was helicoptered out and transported to Australia. The fighter was most recently under restoration at the Historic Aircraft Restoration Society in New South Wales, Australia.

Jungles and swamps still hold a great deal of warbirds, many with extensive combat history. What jungle-find will emerge next from the restoration shops to take to the air?

Thunderbolt in a Swamp

This photograph from the crash report shows P-47B 41-5920 on its belly in a swamp outside of Wilmington, North Carolina. The photo was taken on January 6, 1944, the morning after 2nd Lt. Wesley A. Murphey Jr. belly-landed the fighter. *U.S. Army Air Forces*

Two P-47 Thunderbolts departed Bluethenthal Field, Wilmington, North Carolina, on January 5, 1944, at 5 p.m., local time. The Thunderbolts were part of the 340th Fighter Squadron, 348th Fighter Group based at Bradley Field, Connecticut. In the cockpit of the lead aircraft was 2nd Lt. Wesley A. Murphey Jr., followed by his wingman, 2nd Lt. Virgil D. Newby. The two were en route to Fort Myers, Florida.

Within ten minutes of taking off, Murphey's Thunderbolt developed problems, as its landing gear would not remain in the wells. Murphey cycled the gear a couple of times, but the right main gear would not stay retracted.

Wingman Newby then saw flames stream from the underside of Murphey's P-47B, serial number 41-5920. The flames changed to thick, black smoke as Murphey turned the huge fighter back toward Bluethenthal Field. As he rolled wings-level, Murphey's problems multiplied when the automatic propeller control seized. Murphey was able to manually control the prop, but not for long. Unable to maintain power or altitude, at five hundred feet Murphey made the decision to ride the fighter in as he was too low to bail out.

Murphey brought the fighter down, wheels up, sliding across the slick surface of North Carolina's Green Swamp with the Thunderbolt rotating horizontally on its belly as it came to a stop. Meanwhile, overhead, Newby was radioing for help and once he saw Murphey was relatively okay, he flew the twenty-five miles back to base.

Assessing his situation, Murphey knew he could not walk out from the crash site in the dark as he was dressed for a night on the town in new slacks and shoes, so he hunkered down to spend a cold night in the swamp. The following afternoon, a search party reached Murphey and the Thunderbolt, and he spent the remaining hours of daylight walking out of the swamp.

P-47B 41-5920 was later stripped and abandoned to the swamp, while Lt. Murphey went on to fly Mustangs with the 457th Fighter Squadron of the 506th Fighter Group, 7th Fighter Command stationed at North Field (Airfield Number 3) on Iwo Jima. His Mustang, P-51D-20-NA 44-63291, was known as *Nip Nocker*. Arriving in the Pacific in May 1945, Murphey didn't see many Japanese aircraft in the air; however, he still managed to acquire one confirmed aerial victory and one damaged.

Thunderbolt 41-5920's New Life

The oldest surviving P-47 Thunderbolt, P-47B 41-5920, the thirtieth off the assembly line, is being restored by Randy Ferris of Marengo, Illinois. Ferris and his son are in the crop dusting business, and Randy's father was a B-17 pilot in World War II and a flight instructor in the post-war years. Randy was raised in and around aviation, and while bringing up his own family, he was pursuing rumors of a downed P-47 Thunderbolt. Eventually he found the Thunderbolt and the landowner was more than willing to give him permission to recover the rare fighter. The landowner had been willing to have the fighter removed from the property in the past, but everyone who tried gave up when they saw the logistical hurdles that had to be overcome to get the aircraft home. As Ferris was receiving permission, he was wished the best of luck and told it could not be done.

Don't tell Randy Ferris that he can't do something.

He began collecting P-47 parts years ago, and through his pursuit he met Mike Stowe who runs Accident-Report.com. Stowe and Ferris purchased all of the aircraft crash reports on microfilm from the very early days through approximately 1955. Stowe has been databasing the reports, "and he's really knowledgeable about the crash sites and what's out there," said Ferris. "I told him I'd like to find a Thunderbolt somewhere and try to recover it. Then restore it to flying condition. He said, 'Do you know there's one in Green Swamp, North Carolina?' Over the next couple of years we went down to North Carolina a number of times. We rented an airplane and located the wreck out in the swamp."

The fighter sat on land owned by a nature conservancy. At one time it was owned by a large paper company who had gone in and harvested most of the trees. When the land became unproductive, it was donated to the North Carolina Nature Conservancy. "I wrote letters to the conservancy and the person who was in charge of it," Ferris said. "She gave me a letter of permission to go back in there and do what we wanted to do. One of the stipulations was that we could use the fire roads, but we could not drive off-road or drive heavy equipment or vehicles out to the aircraft crash site. I replied that we were probably going to use a helicopter for the recovery and that sealed the deal."

Driving from Illinois, Ferris—sometimes alone and sometimes with his family—and a group of friends, would leave home on a Wednesday night headed for North Carolina. They'd drive all night so they could spend a couple of days at the site. "We'd walk into the swamp in the morning and disassemble pieces of the fighter, work on it, dig it up, and dig it out. Then we'd go back to the hotel at night and come back the next morning and kept working," said Ferris. "I think we made about a dozen trips down there.

"On the first couple trips, I had a bunch of guys with hatchets and axes and we actually had to cut a path to the wreck. The first hundred yards was like cutting a path through impenetrable foliage. It seemed like we cut about a couple hundred-foot tunnels through this stuff, just to get through it and back into the swamp area.

"The actual walk back to it was like you were walking over waves of undergrowth. It was quite an ordeal to walk into the crash site every day. It was about a mile and one-half hike, and it wasn't like a watery

Aerial view of P-47B 41-5920 before the recovery began. Although the ground looks firm from the air, Randy Ferris said it was like walking on one large plant from the road all the way to crash site. *Randy Ferris*

swamp, like you were walking through water. It was a deep, boggy swamp. You were actually walking on what seemed like one continuous plant. You had to hack your way through all sorts of underbrush and it was really quite a project just to get yourself back in to the airplane. I carried in shovels and axes, torches, and all kinds of tools so I could get the airplane apart."

After a couple weekends' worth of work, Ferris began to store shovels and axes and other tools inside and around the wreck. On a return visit, he quickly discovered that some of the locals had used his trail to the wreck because all of those tools had disappeared. Through the years, the locals had gone to the wreck and rooted around. "They'd wrestle some parts off the airplane and scratch their names or initials into the metal," said Ferris. "In talking with a local pig farmer, he told

A series of fire roads crisscrossed the swamp and it was on this patch of hard ground that the trucks and trailers were staged. A helicopter was brought in to lift the wreckage from the site to the road where it was loaded for the trip back to Illinois. *Randy Ferris*

me they'd been into the wreck shortly after it happened and moved the handles in the cockpit.

"From what he told me, the Army took the guns and ammunition out along with any sensitive instruments. At the time, it was a pretty high-tech fighter, and after they disassembled what they wanted to out of it, they blew up the fuselage and they dropped a couple grenades in the gun bays of the wings. The pig farmer said his kids would go in and take parts off the fighter every once in a while. I asked him if he had any of the parts. He said he couldn't remember what happened to them or where they were. That was disappointing.

"I also met a farmer who said that when he was a boy, about ten years old, he watched the pilot having trouble with the engine. He actually watched him go in and his grandfather helped the military get back in there to get the pilot out the next day. There are lots of interesting stories connected with this airplane."

After a dozen trips to dismantle the P-47, Ferris had made arrangements with a helicopter company to come pick up the parts. They had the parts ready to sling load out, and the helicopter arrived early in the morning, landing on the fire road. After a short briefing, the airlift started. "The engine was the first piece lifted out of the swamp and as soon as the helicopter picked that thing up, that was another unbelievable feeling seeing the parts starting to come out of the swamp," said Ferris.

After five helicopter loads, Ferris's friends began loading the disassembled Thunderbolt onto waiting trucks and trailers. As the loading progressed, "one of the worst thunderstorms I've ever been in came through," said Ferris. "Just a torrential downpour. We were on these fire roads that were dry one minute, but quickly became soaked. About a half mile before the exit out of there, there was an uphill climb and, unfortunately, one of our trucks got stuck there. So we spent the next four hours pushing and pulling and winching and digging and doing everything we could to get the truck up that slimy, slippery road to get out of there. We didn't get out of the swamp until midnight that night.

"Then we had another little incident going back to the hotel in Wilmington when one of the guys cut a turn a little close and took out the rear of another guy's pick-up truck. There's always little things that slow you down, and that made for a long day."

Ferris got a fairly complete set of wings, a damaged engine that will serve as a core, and parts and pieces of the fuselage and tail section. Through the years, forest fires had come through the swamp area and burned the trees, doing some major damage to the fuselage.

"While we were hauling the aircraft out, the woman who ran the Nature Conservancy arrived. We had all different parts laying out on the fire road," said Ferris. "She got out of the car and was absolutely in awe. She couldn't believe that somebody actually had done it, getting that airplane out of there. Yes, she said there were numerous people that had written her and asked permission and she'd given most people permission to go in and do the same thing, but nobody had ever followed through. It was kind of a neat deal to see her reaction."

Looking for Thunderbolts in Every Corner of the Globe

Ferris has been to many of the former battlegrounds in the Pacific looking for Thunderbolt wrecks as well. Not many people realize the extensive use of the P-47 in the Southwest Pacific, especially in the middle years of the war when the tide was turning in the Allies' favor.

A friend of Ferris's took him to a crash site on Australia's Cape York where two P-47s had landed on the beach after running out of gas. Both made successful wheels-down landings and were reported as intact. "For whatever reason, the airplanes were never recovered," Ferris said. "Unfortunately, in the 1970s, the Australian Army went up there and pulled the ammunition and guns out of them and dynamited the hulks."

In addition to Australia, Ferris scoured the Solomon Islands looking for wrecks. He had talked to most of the people in the government and those in charge of the national museums, and came up with an agreement using their generic schoolhouse plan. On whatever island he found recoverable aircraft, Ferris would build schoolhouses for the locals in exchange for an export permit.

With the deal in place, Ferris returned to the United States in search of investors. While trying to run a business and raise a family, things fell by the wayside, and he found it extremely difficult to run a second, aircraft recovery business halfway around the world. "If I were twenty-five years old and not married, no kids, and no other obligations, I'd have been over on the Solomon Islands working that place," he said. "There's so much materiel over there. You could do really well if you wanted to.

"The problem is that the islands are not like a third-world country; they are a fifth-world country. We live over here with our cellphones and computers and TVs and radios and everything, and you go over there and literally you're out in the middle of nowhere. There's nothing," he said. "The way they communicate is every couple weeks an airplane flies into one of the old military airstrips and takes people out, goes here and goes there. They get around and do things, but on some of these islands there's no electricity, there's no nothing. If you get a hankering for an ice cream cone or something, you can't go down to the Tastee Freeze. Every day you never knew where your next meal was coming from and trying to get decent water to drink was difficult. It's a very different lifestyle.

"By the time I got home I had lost about twenty-five pounds. It took another two weeks to stop losing weight. In total, after about

Randy Ferris pauses for a moment while preparing the engine and propeller to be sling-loaded from the crash site. Although the lower section of the engine suffered from corrosion, having sat in the swamp for decades, the upper portion has yielded a number of parts that will be useful during the fighter's restoration. *Randy Ferris Collection*

three or four weeks, being gone a couple weeks and then getting home, I was down about thirty-five pounds. It's not something where you take the wife and kids on adventures like these," he cautioned.

One to Make Six

With the help of a friend, Ferris has put all of the microfilmed P-47 blueprints on disks and indexed them. When he needs a print to make a part, he goes to the computer and pulls up the needed drawing. Lately he's been working on the elevator, rudder, and horizontal and

vertical stabilizers. "I've been missing a bunch of parts, and when I set something up, I'll make enough parts for six airplanes," he said.

"Everything I took out of the swamp is here in my workshop. Over the years, I've been manufacturing various parts and bits and pieces. I've got the engine and propeller here and torn apart. The Curtiss electric propeller had steel blades, which are pretty rusty from sitting outside all these years. As the water in the swamp would rise up and go down, the lowest cylinders on the engine corroded off of it. There are, however, some parts on the upper engine that will be serviceable and usable. If at all possible, I'll probably try to put them on the flying airplane."

Ferris is duplicating fuselage parts that he either buys, or borrows from other restoration projects. "My ultimate goal in the next couple years is to be able to spend a little more time working on the airplane than I have in the past, making various parts and components for the airplane. If I don't have the parts, I just make whatever I need off the print. During the last fifteen years I've filled shoeboxes and cigar boxes and all sorts of containers with parts, and I have big pieces lying everywhere. I've got some controllers and attach points and control rods and turn tabs.

"I don't have any major components assembled for it yet. I just bought a two hundred–pound press that I'm setting up to make ribs and some of the bulkhead components, and things like that. Over the next couple of years I'm hoping to start assembling components that begin to look like an airplane so that people can see that I've been doing something. You can work on a project for ten years and outwardly it doesn't look like you've made any progress, then once you start riveting and putting things together, it starts to take shape. I'm nearing that stage.

"If I had bought a Mustang or maybe a Corsair project I probably could have gotten a few more people interested in it. But nobody's really interested in Thunderbolts. This is something l like to do and I want to do," Ferris said. "I enjoy working on this airplane."

Opposite: The engine and propeller depart the crash site by helicopter. The helicopter loaded these parts directly onto one of the trucks. *Randy Ferris*

Following pages: The helicopter begins to lift the both wing panels from the crash site. At this point there was still a substantial amount of smaller parts to be recovered. *Randy Ferris*

Sole Surviving
Brewster F3A-1 Corsair

At some point, a point that is now lost to history, the Brewster Corsair got a bum rap. During World War II, Brewster had its share of labor troubles, compounded by a series of aircraft designs that underperformed. Thus, the company never reached its full potential. Caught up in a guilt-by-association reputation were the company's license-built Corsairs.

In 1940, Vought's F4U Corsair design was one of the hottest things with wings. Vought licensed production to the Brewster Aeronautical Corp. in November 1941, and to the Goodyear Aircraft Corp. one month later. Brewster built 735 Corsairs, with half of that production going to the Royal Navy's Fleet Air Arm. The U.S. Navy closed Brewster's production line at the end of June 1944, because the company was continually behind schedule delivering the much-needed Corsairs. The Navy shuttered Brewster because of labor issues, not for how its Corsairs performed.

Brewster production and engineering test pilot Ralph O. Romaine said: "From the feedback we received from the Aircraft Delivery Units of the U.S. Navy, the Brewster Corsairs were considered of very high quality and trouble-free. History was never very kind to Brewster for some reason. At the time of contract termination, for the convenience of the government, it was stated in the media that the unit cost [per Corsair] at Brewster was excessive to either Chance Vought or Goodyear. I have always found this hard to believe. The total employment at Brewster was a bit less than four thousand and they were producing about one hundred Corsairs a month. Later I learned that Goodyear in Akron had in excess of twelve thousand workers and a production rate, in the same period, of 150 Corsairs a month."

In spite of what propaganda of the period tells us, there were labor issues and strikes across the United States during World War II that impacted not only Brewster, but other aircraft factories and shipyards as well. Each strike had an effect on war production, and Brewster as a company was terminally impacted by this unrest.

The Corsair's reputation should be intact, no matter which company built the individual aircraft. Brewster's production was the smallest, and half of those built went to the Royal Navy. The Brewster Corsair population was further reduced by those involved in accidents, and then subtracting those scrapped after the war, it should be

Brewster test pilot Ralph Romaine pulls an F3A-1 in close for an air-to-air portrait. Somehow the rumor got started that the Brewster Corsairs were sub-par, but test pilot Romaine dismissed such notions. *Ralph Romaine via A. Kevin Grantham*

no surprise that only one Brewster F3A Corsair survives. However, it survives because it was hidden in a swamp for decades.

Sole Survivor

The sole surviving F3A-1, BuNo 04634, was accepted by the U.S. Navy on January 8, 1944. Interestingly, the record card shows an assignment to "PAC" on January 28, 1944, and its next assignment, to VMF-124, takes place seven months later in August 1944. One has to wonder if this indicates an overseas combat assignment? The Corsair was then assigned to Marine Air Wing (MAW) Nine, then Headquarters, MAW Ninety-Two, and on December 8, 1944, it was assigned to a Marine Fighter Squadron (whose number can't be discerned from the handwriting on the record card), then on December 23 to Service

F3A-1 BuNo 04515 was the prototype Brewster-built Corsair. Notice the short leg tail gear making the airplane sit extremely nose-high. On production aircraft, the tail gear leg was extended to lessen the problem. With the nose sitting so high, it is obvious how the forward fuselage and the front of the wings drastically reduced forward visibility, especially when trying to land the aircraft aboard ship. *Ralph Romaine via A. Kevin Grantham*

Squadron Ninety-One. In January, February, and March 1945, BuNo 04634 was reported as part of the aircraft pool at Marine Corps Air Station Cherry Point, North Carolina. This reporting is probably in error because 2nd Lt. Robin C. Pennington perished when he bailed out of the aircraft on December 19, 1944, and the accident report most likely had not caught up with the individual aircraft record card to reflect its destruction.

During a ground-controlled intercept flight ten miles southwest of Cherry Point, Lt. Pennington was to make a run on a Marine Corps

PBJ (B-25 Mitchell) at 3:00 p.m. When he failed to meet up with the PBJ, a search was launched for his aircraft. It was found north of the hamlet of Great Lakes, North Carolina.

It appeared that Pennington had bailed out of his aircraft as the seat belts were disconnected and the canopy was jammed open about four inches. He may have been struck by the fuselage or tail of the Corsair when he went over the side as his body was found in a tree some distance from the aircraft.

Pilotless, the fighter continued to fly, setting itself down in a swamp at a shallow angle. In the transition from flying craft to scrap metal pile, the left wing struck first and was torn from the fuselage, and the engine and its accessory section broke loose from the right side of the fuselage. Heavily damaged, but certainly rebuildable, the Corsair had landed in a swampy peat bog.

For more than forty-five years, the Corsair sat relatively unmolested in a North Carolina swamp.

The stripped remains of F3A-1 BuNo 04634 as found by Lex Crawley sitting in a swamp near the Marine Corps air base at Cherry Point, North Carolina. *Lex Crawley*

Enter Pirate Lex Crawley

Lex Crawley was bitten by the warbird bug at a young age. Friends say an airplane flew over the family farm and he was hooked from then on. At some point he saw photos of German planes at the end of the war in varying states of disrepair and that started him digging a series of deep holes on his family's farm in search of buried Axis aircraft. He didn't find anything on the farm, but instead of being discouraged, he was only more enthused to continue his pursuit.

"I was about fifteen when I found out they buried government surplus aircraft here in the United States at a local university in the area where I lived. I pestered the airport manager for weeks until they finally let me go out and dig up a little bit of the T-33, T-6, and Beech 18, so the bug was firmly planted," said Crawley.

In the mid-1970s, Crawley became an aviation insurance underwriter and broker in Southern Illinois and began talking to people about salvaging warbirds. "I was surprised there were quite a few wrecks out there, and in 1979, I went to Cleveland and flew my first air race in a Formula V. Just by chance, I dropped in on a little airport to buy gas and met Walden 'Moon' Spillers, who showed me his P-51 project (P-51A 43-6006). After seeing his project, I had to find a wreck of my own—somewhere. The search was on."

Through contacts in the aviation insurance business, Crawley learned about a Thunderbolt in a swamp in North Carolina. It was in a bridge area, and in much worse condition than he had hoped, but it did exist. Then he heard about a Corsair in a swamp, but people told him it was a pile of junk because the control stick was gone and it had a broken wing spar. "In 1979 'warbird dollars,' it wasn't worth the price of getting it out," he said. "I went out and found a number of other warbird wrecks over the years. I had located a number of Mustangs, P-40s, a couple of Thunderbolt wrecks, and, about 1990 I was talking to Conrad Eggan here in Minneapolis. He mentioned the Corsair wreck and I asked if it was still in the swamp. It was, and I asked him if he had any interest in going after it. Eggan said, 'I have no interest in getting out there with the snakes and the cottonmouths to try and recover anything.' I immediately hopped on a plane, went down and looked at it. I thought, 'By golly it's worth getting.' I didn't even know it was an F3A-1 until just before I pulled it out."

A dozen years prior to pulling out the Corsair, the Navy had threatened to sue Crawley and throw him in jail for trying to steal one of their warbirds. "It turned out to be an Army Air Corps airplane, but at that time I learned that with the Navy it's not a matter of paperwork first, it's a matter of possession first. When the shoe was on the other foot and I had found the Corsair, I thought to myself, possession is nine-tenths of the law and I felt it was legal to recover it. The aircraft sat on private property after it crashed and was abandoned, and the landowner gave permission for me to remove it. We worked on her for two days and then used a Sikorsky S-58 helicopter to lift the aircraft out of the swamp."

The Corsair was sitting in what was essentially a peat bog, and using a helicopter was the fastest, and safest way to recover the fighter. "We loaded her right as she was. Actually, we had about 2,500 pounds there at the best calculation," Crawley said. "The helicopter operator had a digital scale connected to its hoist hook-up and he said I was within a couple of hundred pounds of the actual weight. They took out as much as I could put into the fuselage. I think 3,500 pounds was their actual maximum lift. I figured I stayed under 3,000, maybe 2,900 pounds. They were shocked that we were that close in our

The Corsair's main fuel tank sat in the forward bay, and the cockpit was in the rear. Through the years, most of the F3A's major components were removed by the Marines and area locals. *Lex Crawley*

estimate because they said, 'Everybody says it's only X pounds and it never is.' On our second load we brought out the left chunk of the center section and left outer wing panel stub in one lift."

Moon Spillers taught Crawley to pick up everything at the crash site, airworthy or not. Anything that had the slightest chance of being a pattern would prove useful later on. The Corsair was lifted out of the swamp on January 18, 1991.

The Corsair was trucked to Crawley's father's farm in Georgia, and it sat there for three years. At the time, Crawley was a mechanic with Northwest Airlines and was moving around the country fairly regularly. When he settled in Minneapolis in 1994, the Corsair came north.

View of the F3A's fuselage looking to the rear toward the tail. There is a trim wheel on the left side cockpit wall (to the right in the photo). *Lex Crawley*

Under Attack from the Naval Historical Center

Before proceeding, it should be made clear to the reader that Lex Crawley's trouble with the Navy stemmed from the Naval Historical Center in Washington, D.C., not the National Naval Aviation Museum in Pensacola, Florida. And although their missions seem similar—to preserve the Navy's history and artifacts—they operate very differently. "Don't let it be misunderstood: the National Naval Aviation Museum is a great institution and a great group of people and volunteers. I'd do anything to help them," Crawley said.

Oddly enough, Crawley's problems with the Naval Historical Center began with a local Minneapolis man who kept trying to buy the Corsair project. He repeatedly made low-ball offers and Crawley refused each one. Seeing that he could not acquire the Corsair, this gentleman began telling people in the warbird community that he had bought the fighter. After a confrontation with the gentleman, Crawley phoned the Naval Museum to see if he could trade the ultra-rare Corsair for one of the fighters recovered from Lake Michigan—anything to keep the Minneapolis man from getting ahold of the Brewster fighter. "I knew right then and there, either I'd make an exchange, or the Navy would throw me in jail and take the airplane, but it would never end up in this individual's hands," Crawley said.

When the trade proposition was moved up the chain, out of the museum's hands and into those of the Naval Historical Center, things started to get weird. Questions started coming from Washington that were clearly fishing for answers that could be used against Crawley in one way or another. This went on for five years. They made suggestions to Crawley such as, "Why don't you put it in a business name so it looks more correct us dealing with you?" However, he was not a business, nor was he in the warbird recovery business. "They were trying to bait me in so many different ways to spin potential outcomes that would look much better for them in publicity when they came after me," Crawley said.

Before things got really bad, a former U.S. senator that Crawley casually knew viewed all of the information, and said, "You're absolutely kidding me. These pictures of the airplane in a swamp are junk, and they want it?" The former senator went to Washington, D.C., on his own time and money to talk to the Undersecretary of the Navy. Soon after that visit, a lawsuit was filed against Crawley in 2003.

The interesting part of the story is that Crawley had registered the aircraft with its true identity a decade earlier, and nothing was said. With it being the only Brewster aircraft on the registry, there was no chance Crawley's Corsair would impact the title of another owner and he certainly was not hiding the fact that he had recovered the fighter.

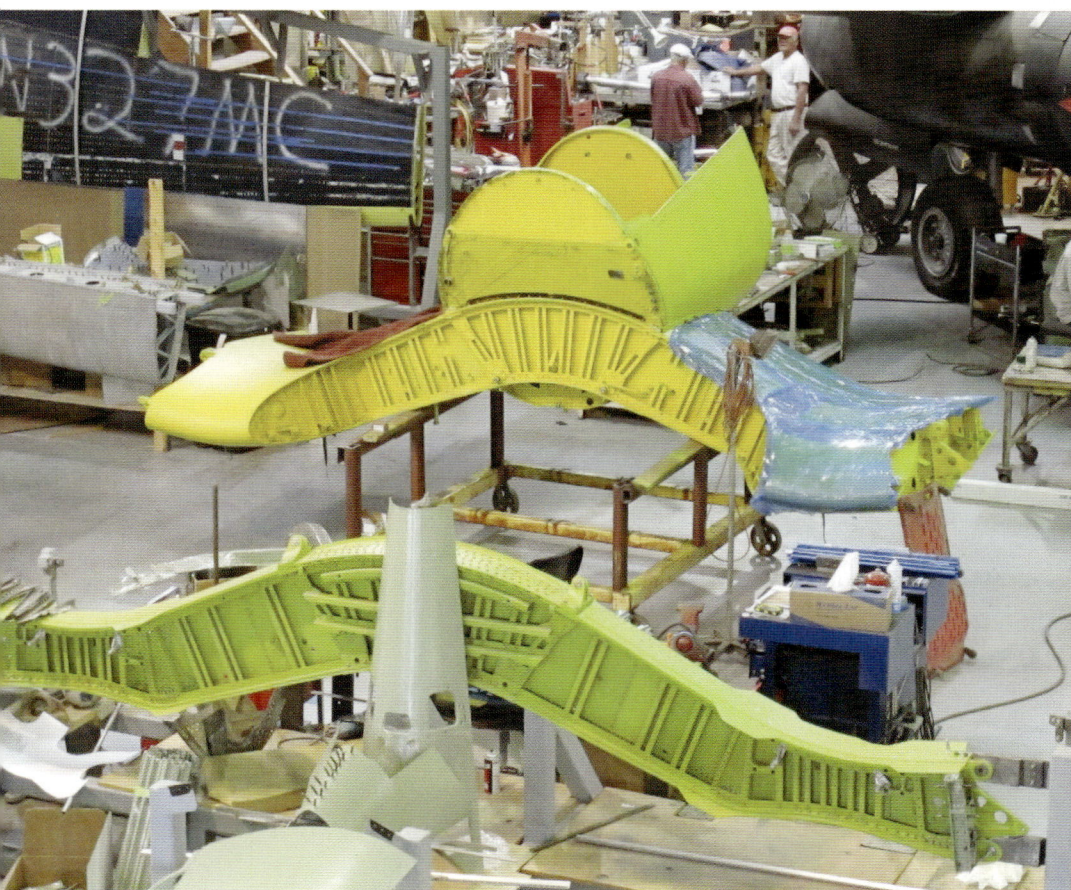

Early in the restoration stage, the Brewster F3A-1, foreground, sits in its jig with the Commemorative Air Force's Goodyear FG-1 Corsair center section behind. *Robert Nishimura*

The F3A-1's main spar and stub wing ends sit in the jig, attached at the wing-fold castings. *Robert Nishimura*

"In every exchange of information with the Navy they would ask me for something. I would make an offer or counter-offer and it took invariably six months for it to go through the channels, from desk-to-desk and attorney-to-attorney. It usually turned into a letter and returned to me as a new threat," Crawley said. "In the end, I came home one morning after working a midnight shift and my wife looked at me and said there had been a reporter calling from the *Pioneer Press*, which is the major newspaper of St. Paul. The reporter said he wanted to ask about the lawsuit. I didn't know anything about it. It wasn't until later in the day I received a written notice by standard delivery mail. It wasn't even certified mail, or courier, or anything. The Naval Historical Center tipped off the *Pioneer Press* reporter before I'd seen the paper, and the reporter then started calling. I'm sure everyone at the Naval Historical Center was surprised when their tip blew up in their faces.

"It's fortunate that I did agree to talk to the reporter. I was concerned because my attorney was out of the country at the time. This reporter seemed to be very fair, very interested in the facts, and he wrote a story that literally, once it hit the Associated Press, went global. Then the phones started to ring. Then it was the front page, lead story on Saturday morning. Obviously it was a slow news day.

"A week later, politicians were starting to call, and a big player in the story turned out to be Congressman Walter Jones from North

Carolina. Thank goodness for caller ID at the time; it was plugged into our TV and it said 'U.S. Capitol' or otherwise I would have thought it was a prank call. Congressman Jones asked if I was Lex Crawley and I said, 'Yes sir.' He identified himself and he asked, 'Are you the one that has this Corsair?' I could tell he was irritated. I thought, 'Goodness gracious! I got this Corsair from his district and now he's feeling I stole it from him and his constituents.' I thought, 'I'm in even more trouble.' Then after he talked a couple of minutes, I realized, 'Wait a minute! He's mad at the Navy!' He ended up being just a wonderful person, a huge help. It would have turned out completely different had he not been involved. More than likely, it would have gone to trial and it is anybody's guess how that would have turned out. He was absolutely instrumental in dealing with the Navy. I listened in on some of the phone calls and he didn't pull any punches.

"And I'm sure Congressman Sam Graves went to bat for me as well. He was upset that they were just going to leave it there to let it rot. He was also upset that they had abandoned it for fifty-six years and then they're trying to say it's theirs. He thought what an absolute waste of taxpayer money! He wanted to know why I was being harassed and sued. Once he saw photos of the airplane, he was even madder. He said, 'That is just a pile of junk! You're trying to preserve it. You're trying to save history. They had no intention of getting it and they want to harass you and sue you for it?' He's a very practical, conservative man. I've gotten to know him over the years since then. I have really high regards for him and what he did for me."

With political intervention between the Naval Historical Center and Crawley, ownership of the Brewster Corsair was transferred to Crawley through a provision in the 2005 National Defense Authorization Act. Once the act was signed into law, the Corsair belonged to Crawley, free and clear. Summing up his battle with the Naval Historical Center, Crawley said, "They don't understand that the airplanes are deteriorating very rapidly. Many of them have been lost through the center's policy of letting the aircraft remain stored *in situ*."

Soon after the title was transferred, Crawley began restoring the Brewster Corsair. At about the same time, warbird enthusiast Jim Slattery began building his impressive collection of U.S. Navy aircraft. A deal was reached between Crawley and Slattery and the Brewster Corsair was shipped to Ezell Aviation in Breckenridge, Texas, where it is now undergoing restoration to flying condition.

Returning to Fly

The Corsair project arrived at Ezell Aviation in February 2011. With it came the marching orders to return it back to stock condition while trying to use as many of the original parts as possible.

The next step was to inventory what had arrived. There was nothing firewall forward; the cockpit had been stripped except for the left side console; the right wing was present because it had remained on the aircraft during the crash; while the left wing had struck the ground first and absorbed the impact of the crash and ensuing cartwheel.

"The whole left side of that airframe was pretty much destroyed as far as the wing and center section," said Chad Ezell of Ezell Aviation. "There were chunks of the wing that was attached to the center section, which Lex Crawley had removed with a torch to make it easier to move. It really didn't hurt anything because these parts were not salvageable.

"The worse thing about what the aircraft went through was sitting in the swamp. The ribs were good on the whole bottom of the center section, but what sat in the mud on the right side corroded. Because it sat in a swamp, combined with the acidic soil, the lower structure was in poor condition. When we started pulling it apart, the landing gear was just one bonded unit of corrosion and rust and we had to cut a lot of that out. Once we started getting into the rib structure, they appeared to be in good condition until you got to the bottom where they were severely corroded and pitted.

"The cockpit fuselage sections had problems because they were caved in during the landing impact. There were several ripples through the sides and we'd find one useable rib out of every four. And then you had bullet holes in what might have been useable parts."

In an effort to separate what was useable from what could be used as a pattern and what was scrap, everything was tagged, photographed, and logged into Ezell Aviation's computer system.

Recently the center section was mated to the fuselage to prepare for building the rear spar sections. The corrosion on the right side of the aircraft and the missing part of the left side required that all new spars be made for the rear section.

"The spar section, which we didn't have and we were looking to engineer and build ourselves, was acquired from Duane Doyle," Ezell said. "He had already been down that road and had built four complete spar assemblies. He was willing to sell us one because he'd decided he didn't want to build that many Corsairs because they're too

An original fuselage skin panel from the F3A-1 is lined up on the restored rear fuselage. Notice that part numbers have been written on the exterior of the skin to help technicians reference stringers on the inside of the fuselage. Every part Vought installed into the aircraft had a part number, and Brewster and Goodyear continued this practice. In comparison, Grumman did not maintain this practice with the F6F Hellcat or F8F Bearcat. *Robert Nishimura*

complicated. Buying that one spar assembly saved us a year's worth of work because we would have had to work with an engineer out in California who's made all of these parts in the past. That helped us out and we put that into the jig and just had to build everything around it.

"We had some tooling to make the leading edge skins from doing other Corsair work for Jim Fryer who owns Howard Pardue's old Corsair wreck. To make the inner ribs, we just tooled-up to make new ones. There's actually part of that left gear section where a lot of those complicated ribs are still intact. Pretty much the whole center section is about ninety-seven percent new airframe. I know the spar is a hundred percent new because we bought it, and there's some small doublers and small stringers, and some parts throughout that I'd say were reusable. But I'm having a hard time counting up to make a good handful. They're also working on the rear spar sections right now, which involves making new flap-wells. There are two panels on top and bottom of the center section outside the fuselage that go on as assemblies, and those are spot welded together but you've got to build up the rear flap-well before you can attach all that to the rear spar. So that's in process right now."

Windscreen and glare shield of the Brewster Corsair seen after restoration in the Ezell shops. As each subassembly is restored, it is primed before mating. The aircraft's rear fuselage sits in the background. *Robert Nishimura*

The fuselage had a number of parts that could be used as templates, many of them ruined by bullet holes. In addition, the fuselage had a twist to it due to the impact, which kinked everything. The Ezell crew was able to save some ribs, but ended up having to make many new ones for the mid-section of the fuselage.

Looking at the aircraft's structure and comparing it to other Corsairs they've had in the shop, Chad Ezell estimates that BuNo 04634 was a mid-stream production aircraft. However, with all of the missing pieces, there's a lot of finishing details that cannot be discerned.

Interior and Exterior Paint

There was recently a Corsair recovery that showed the aircraft had been built using a red primer. The Navy's specifications call for two coats of Chromate paint, and each coat had to be a different color to prove that all areas were being covered. The Corsairs used a lot more Chromate yellow as a top-coat while the base coat was a Chromate lime. Brewster liked to use a salmon color in its aircraft.

The area forward of the firewall was typically painted green while the wings were a mix of colors. Some wings have had a light green primer under a dark green top-coat, while the Brewster Corsair's left wing had a light green primer with red top-coat.

When deciding the exterior color scheme for BuNo 04634, owner Jim Slattery met with restorer Chuck Wahl of Vulture's Row Aviation in Cameron Park, California, and the Ezells and decided to put the Brewster Corsair back to its original paint scheme. It had a tri-color blue scheme with an L-69 buzz number on the sides of the fuselage.

Wing Details

The Corsair is built like a tank, and what would be a simple structure on another aircraft is extremely over-built on the Vought design. "I'm still baffled, as is everybody in the shop, on how complicated Vought made a simple structure," said Ezell.

When the Corsair project arrived in Texas, the right wing was intact, and the left wing was in pretty sad shape. Duane Doyle, who's rebuilding a Corsair in the San Francisco Bay Area, had some early model Brewster wings, which he sold to the Ezells.

"The wing we acquired from Duane Doyle had fuel tanks in the leading edge, so we know it's an early model wing," said Ezell. "Right now we've got the left wing in the jig and are ready to apply the base primer for assembly. Then we'll install the fuel cell, which is really just a wet wing.

"If want to talk about another nightmare of assembly, it is the jigsaw puzzle of the wing. I don't know why they couldn't just make a wing. All it really is, is a spar, and then you've got a few ribs, and then you've got a gun bay, and they overcomplicated that beyond what you could imagine. How they could come in and have fasteners overlap fasteners I'll never know. This means if you don't install pieces in the correct order, you've got to start over—you've got to pull stuff back out because certain assemblies can only go in one way. Then it gets overlapped with another fastener and then they'll put stuff in and then they'll rivet it, and then they add more fasteners around it.

"The mindset of the engineers in the late 1930s is something to marvel at today. My guess is that they were using a brand new material, aluminum, and the mindset was very much like our engineers using composites now. The 1930s aeronautical engineers had been building airplanes with steel frames covered in fabric, and before that it was wood covered in fabric. Aluminum was lighter and they were scared to death of it. They truly had no idea of just how strong this lighter material was. To combat their fears, they'd add another rib, or put in a casting. That's all I can figure because the wings are way overbuilt. I think the leading edge skin is 0.63 inches which is really thick for lead edge skin. Then in the center section the skin is 0.80 inches—very thick. Of course a lot of it was because that the structure had to hold together while bouncing onto a carrier deck."

Halfway Point

The sole surviving Brewster Corsair has done battle with the elements in the swamp as well as the bureaucrats in the Naval Historical Center. It is now three years into a five-year restoration, one where the entire aircraft has been stripped down to every last nut, bolt, and rivet, and a lot of it is being manufactured from scratch. It will emerge as one of the finest Corsair restorations ever under taken.

Recovering the Sandbar Mitchell

In 1969, wildfires burned more than 4.2 million acres in Alaska, and North American B-25 Mitchell bombers were there to battle the fires. Air tanker operators would buy surplus Mitchells, remove all of the bomber's military equipment and install a 1,000-gallon tank in the bomb bay. At the time, air tankers were dropping Borate retardant on fires and 1,000 gallons weighed approximately 11,000 pounds, pushing the B-25 near its maximum gross take off weight.

On June 27, 1969, at approximately 2:30 p.m., pilot Herm Gallaher had just lifted off from the airport at Fairbanks, Alaska, in TB-25N N9088Z, also known as air tanker Eight-Zulu. The military surplus twin-engine bomber was hauling a load of retardant to be dropped on the Manley Hot Springs fire, ninety miles to the west.

It was sunny and around seventy degrees that day as Gallaher retracted the air tanker's landing gear. While establishing his climb, both engines suddenly lost power.

Gallaher quickly dropped his load of retardant and with very little altitude beneath him, he began looking for a place to set the Mitchell bomber down. In front of the Mitchell's nose was a sandbar on the Tanana River, about three and one-half miles south of Fairbanks. The sandbar was devoid of trees and only covered with scrub brush. It was good a place as any to belly-land the bomber. Gallaher skillfully set the bomber down with only the underside of the aircraft and the right outer wing panel suffering much damage. Gallaher walked away from the incident to fly another day and continued his long career as an air tanker pilot.

A few days later, aircraft owner Edgar Thorsrud surveyed the bomber. The cost of salvaging the aircraft exceeded the price of a used B-25, and the decision was made to abandon the bomber. Thorsrud sent his mechanic, Jim Andersen, out to remove the two R-2600 engines, strip the cockpit instruments, tires, wheels and brakes, and

N9088Z sits at the tanker base in Anchorage, Alaska, when operated by Edgar Thorsrud, who owned a fleet of B-25 and TBM air tankers. *Milo Peltzer*

Pilot Herm Gallaher set Eight-Zulu down on a Tanana River sandbar after losing power while climbing out from Fairbanks, Alaska, with a load of retardant on July 27, 1969. Gallaher was en route to drop retardant on the Manley Hot Springs fire, eighty miles to the west. This photo was taken in 1978, and shows the steady growth of trees in the area. *Milo Peltzer*

take some of the control surfaces. And that was it. What Andersen didn't take was left to the elements.

Thorsrud bought an engineless B-25 that was sitting in Anchorage to replace Eight-Zulu. He mounted the R-2600s from Eight-Zulu to the Anchorage airframe, and was back in business.

Eight-Zulu's Career

The derelict Mitchell sitting on the Tanana River sandbar was built at North American Aviation's Kansas City, Kansas-factory, as a B-25J and assigned serial number 44-30733. The bomber was delivered on

Opposite: By 1986, the B-25 was hidden from sight and could only be seen from the air, by flying directly over the bomber. *Milo Peltzer*

February 16, 1945, and with the war winding down, 44-30733 was stored at Garden City Army Air Field, Kansas, for fourteen months. The bomber was then assigned to South Plains AAF, Texas, in April 1946, and then to Pyote Air Force Base, Texas, in August 1947. In November 1948, the Mitchell was assigned to the 3575th Pilot Training Wing at Vance AFB, Oklahoma, and was redesigned a TB-25J.

Approximately six years later, in March 1954, 44-30733 was upgraded by Hayes to TB-25N standards as a navigator training aircraft. At Hayes, the armor plating and upper turret were removed, the front windscreen was changed to a single piece unit with wiper blades, two seats were installed forward of the bomb bay and five more aft, and bomb bay fuel tanks were added, along with updated radios, an upgraded auto pilot, a new oxygen system, and cabin heaters. After the modifications, the bomber was returned to the 3575th.

In November 1957, 44-30733 moved to the 3510th Combat Crew Training Wing at Randolph AFB, Texas. It flew here for six months before being retired to Davis-Monthan AFB, Arizona, in May 1958.

The TB-25N began its civilian career when sold surplus to National Metals in December 1959. The company registered 44-30377 as N9088Z, and quickly sold the Mitchell to Johnson Flying Service in Missoula, Montana. Here it underwent conversion to an air tanker, and it fought fires with Johnson Flying Service until December 1966.

Edgar Thorsrud, also of Missoula, was a smoke jumper prior to World War II, and flew Ford TriMotors and Douglas DC-3s for Johnson Flying Service delivering men and cargo into the backcountry. He was an aerial fire-fighting pioneer, flying some of the first slurry drops onto fires in Johnson's TriMotors. In the late 1950s, Thorsrud struck out on his own, operating a fleet of firefighting Grumman TBMs and B-25 Mitchells. His aircraft could be seen dropping retardant on fires in Alaska, Idaho, and Montana each season. Thorsrud acquired Eight-Zulu in December 1966, and prepared it for the 1967 fire season. The aircraft fought fires without incident until June 27, 1969.

After the fateful day in late June, it sat abandoned on the sandbar. In the ensuing years, trees grew up around the bomber and soon it could only be seen from directly overhead. Eventually it was forgotten by all but a few locals, hidden from the outside world.

The locals, however, used the abandoned bomber as a source of sheet metal. During the winter when the river would freeze over, they'd come with their snow machines and hack parts and pieces from the craft. Eventually they took the entire upper rear of the aircraft, the outer wing panels, and the bomber's nose. All that remained was its center section.

Forty-four years after crash landing on the sandbar, the hulk of Eight-Zulu's fortunes would change.

Enter the Warbirds of Glory Museum

The driving force behind the recovery and restoration of Eight-Zulu is Patrick Mihalek. He is the founder of the Warbirds of Glory Museum in Brighton, Michigan, and he operates a warbird restoration and maintenance business known as Legend of Aces Aviation. In addition, Mihalek's good friend Todd Trainor, director of the Aeronca Aircraft History Museum, also assisted in the recovery of the B-25. Legend of Aces Aviation has rebuilt a number of AT-6 and Yale projects, and a Ryan PT-22, and routinely maintains other warbirds, from Lightnings to Mustangs. He has the experience and the know-how to rebuild a B-25; the question was could he turn this energy from enthusiastic dream to project on the shop floor?

"When I was in college I'd spend evenings looking through the Civil Air Patrol's records of wrecked aircraft and I stumbled across the Sandbar Mitchell," said Mihalek. "My goal was to build a flying museum around a B-25, but I knew it was going to be very expensive and a Mitchell was going to be hard to come by."

Mihalek's big break came in May 2012 when the nose section from an ex-RCAF B-25, that had spent years on static display, became available. The bomber had been seized by an airport authority for nonpayment of storage fees and was sold at a sheriff's auction. Mihalek was able to acquire the B-25 hulk from the gentleman who had bought it at auction. "I started calling it the 'Ramen Noodle B-25' because I bought it on a Ramen Noodle budget, essentially for what a B-25 cost back in the 1960s."

In May 2013, Mihalek and friend Evan Nunn were heading south to pick up the nose section and a number of small parts they had acquired. Driving a rental truck and pulling a sixteen-foot car trailer, plans called for the parts to go in the truck and the nose would ride on the trailer. They knew it would take two trips to get all of the B-25 parts from Oklahoma to Michigan. While driving down on the first trip, to recover the nose section, Mihalek and Nunn were talking about the Sandbar Mitchell, and they stopped at the Kalamazoo Air Zoo to look at the museum's B-25 and to take some measurements.

Taking a helicopter into the site, the Sandbar Mitchell is seen in 1986. Locals have hacked off the bombardier's compartment and what wasn't removed by the aircraft's owner is rapidly disappearing. The large borate tank can be seen in the fuselage section above the bomb bay. *Milo Peltzer*

As they were leaving the museum, the gift shop had a stack of magazines for sale, and on top was the November 1992 issue of *Warbirds International* magazine that had a feature on the Sandbar Mitchell written by Michael O'Leary and Milo Peltzer. "I looked at Evan and I said, 'If this isn't a sign I don't know what is,'" Mihalek recalled. "I immediately bought that magazine."

With the auction acquisition, Mihalek now had the forward and aft fuselage sections of a B-25, but he didn't have the center section that contains the all-important bomb bay, engine nacelles, and main wing spars. "I still knew there was a lot more B-25 needed to make a flying aircraft, but this gave me hope to actually recover the Sandbar Mitchell."

One afternoon, a gentleman with an Aeronca project phoned Trainor looking for some float drawings for his restoration. Interestingly, he was calling from Fairbanks. Patrick overheard the call and immediately asked Trainor to call back, asking the gentleman if he knew anything about the B-25 sitting on the sandbar south of town. He didn't, but he pointed Trainor and Mihalek to the curator of the Fairbanks Museum, who in turn put them in touch with Edgar Thorsrud's mechanic, Jim Andersen. "Through Jim we heard a lot of stories about Edgar and Herm and we ended up getting in contact with the Thorsrud family telling them what I was doing, and what I hoped to accomplish," said Mihalek. "After about three or four weeks the Thorsrud family decided to sell me the remains of the Sandbar Mitchell. For years people had been trying to get ownership of the airplane, either for the registration, or to scrap the remains, or part it out. Edgar didn't want that to happen, and I think he would have liked to see the components of the Mitchell used in a flying restoration." Edgar Thorsrud had passed away on March 27, 2007.

How Much Paperwork Did the Dream Involve?

Once the paperwork transfer was complete, then the fun part started, trying to get permits. The sandbar appeared to be part of the U.S. Army's Fort Wainwright range complex. Mihalek and Trainor were able to make a contact on the base who shepherded them through the process, getting the base commander to grant permission and obtaining all of the necessary paperwork and permits. The base was happy to have what it considered "trash" removed from their property at no cost. When they heard that, Mihalek and Trainor thought that going forward it would be a piece of cake.

As the center section is rigged, a film crew from the TV show *The Restorers* captures the action for an upcoming episode. The plaque placed by the Warbirds of Glory Museum to commemorate the loss and recovery of the Sandbar Mitchell can be seen at the left foreground. *Courtesy Warbirds of Glory Museum*

"Well, we were wrong, because that piece of land used to belong to the Bureau of Land Management and had been turned back over to the State of Alaska," said Mihalek. "Now we had to deal with the Department of Natural Resources, state preservation officers, Fish and Game, and a number of other agencies.

"We got a hold of the Department of Natural Resources and told them what we wanted to do, explained to them the story, and that I bought the aircraft from the last registered owner. They told us we still had to get the salvage rights, recovery rights, and all this other stuff, and it ended up with them coming back saying they don't control that land, that it belongs to the Bureau of Land Management." Neither agency

B-25 Survivors by the Numbers

Operational	41
Static Display	61
Under restoration	9
In Storage	10
Total Complete Airframes	121
Partial Airframes	22
Total Listed Airframes	143

Source: Scott Thompson, Aerovintage.com

knew who owned the sandbar, and as it turned out they both did . . . sort of. That meant that permits would have to be obtained from both agencies, along with permits from the Department of Natural Resources. Then there were permits from Fish and Game as well as the state Historic Preservation Office. Once the permits were obtained, the Department of Natural Resources opened the recovery effort to a thirty-day public comment period. After jumping through all of those hoops, someone could still derail the recovery effort with a negative comment. Fortunately, none were forthcoming and the process proceeded to the next step.

The Mitchell was sitting on the southwest corner of Fort Wainwright's gunnery range, which made the area a danger for unexploded ordnance. When the recovery team got to Alaska, the first thing they did was receive a briefing from unexploded ordnance (UXO) specialists on what to watch out for.

"One of the gentleman that we brought on the recovery team was retired from the military, and he had experience with UXOs and doing UXO sweeps," Mihalek said. "He made a sweep of the working area around the aircraft and path to the riverbank. It was important to stay within this area so no one would step on any UXO. The problem was that no one knew what had been fired in that area in the early years and we didn't want to find a big mortar shell or anything like that."

Recovering the Bomber

After the paperwork, the next biggest obstacle was getting funding for the recovery. "The one thing we did that had never been done before in the warbird community was crowd-funding," said Mihalek. "I went on line and started a Kickstarter campaign with the hopes of raising $20,000. In thirty days we exceeded our goal and made nationwide news." To everyone's surprise, 402 backers pledged $27,154, thirty-five percent above what was hoped for.

Permits in hand and arrangements made, Mihalek and Trainor set off for Alaska with a team of fourteen people. They also got a lot of help from the locals in Fairbanks. The youngest member of the recovery crew was Logan Kucharek, a fifteen-year old who turned wrenches and helped ready the Mitchell for recovery.

"Our original plan was to do this on a 'ramen noodle' budget. We were to travel to Fairbanks on June 22, take a river boat out to the island, disassemble as much of it as we could, remove all that by the riverboat and then leave the center section out there on a skid until January when the river froze," said Mihalek. "Then we could go out there and pull it down the frozen river, which is how they move big heavy equipment in that part of Alaska.

"That plan allowed us to accomplish the recovery within our budget and timeframe. We couldn't afford a big expensive helicopter lift. With my story getting out and with people learning what the Warbirds of Glory Museum was about and what I'm trying to do, it touched a lot of people, not just in aviation. After hearing our story, Construction Helicopters Inc., of Howell, Michigan, contacted us. They are a heavy lift helicopter company and had a contract up in Barrow, Alaska, at the same time that we were going to be in Fairbanks. One of their helicopters was going to be transitioning through Fairbanks for fuel and the company called us a few weeks before we were getting ready to leave. They said they were going to be in the area with a Sikorsky S-92, and that they'd be willing to support us—and they did."

The B-25 recovery team arrived in Fairbanks on June 22, summer in Alaska, when it never gets completely dark at night. The next day they went out to the crash site to prepare for the recovery. The first task was to clear a path from the riverbank to the aircraft and sweep the area for unexploded ordnance.

Through the years, the river had risen and flooded the sandbar a number of times, depositing layers of silt on and inside the bomber. Mihalek engineered a sawhorse system with a four-inch by six-inch beam that enabled the team to jack the aircraft up using four portable, two-ton bottle jacks. The sawhorse system was built in Fairbanks and floated to the site on the second day. Using the system, the team would jack the center section up an inch or two at a time, brace it, reposition the jacks, and repeat until the aircraft was out of the silt.

Once the aircraft was up on the sawhorses, work began to remove the borate tank from inside the bomb bay. The tank was made from three-thirty-seconds aluminum, and it took ten hours for Mihalek to surgically remove it. "I went through thirty-five Sawzall blades," he said. "We had to be very careful doing this because I didn't want to cut into any part of the aircraft. Luckily when they built the tank they installed a plywood liner between the airframe and the tank. As I would cut, I had some safety with that board. I would cut between

A Sikorsky S-92 from Construction Helicopters lifted the center section of the Sandbar Mitchell and transported it to the airport at Fort Wainwright. The lift and flight took a little more than five minutes. *Courtesy Warbirds of Glory Museum*

the formers, and I had a spotter that was always above looking down at the blade to tell me if I was getting close to the airframe. I cut it out about one square foot at a time. When we were done, we had 650 pounds of scrap aluminum from that borate tank."

The remaining part of the lower, aft fuselage had been buried in silt as well. That section was dug out and removed by hand. It was nearly the entire rear fuselage floor including the door and boarding ladder.

On the third day, June 25, the team broke into groups, working on the inboard flaps, the aft nacelle sections, and the landing gear doors. Then the remains of the forward and aft fuselage sections were carefully unbolted. Another group worked on removing hydraulic lines and the control cables, while Mihalek worked in the bomb bay under the aft fuselage removing the aileron bell crank and the emergency flap extension gearbox. Next task was to remove all of the hardware around the attach angles of the fuselage.

The fourth day saw the forward and aft fuselage removed from the center section. The flat-bottom riverboat took all of the parts, including the 650 pounds of borate tank scrap, back to Fairbanks. It could haul 2,500 pounds per trip, and they kept the boat busy all day.

On the day of the actual anniversary, June 27, nearly everything was off the island except the center section and the remaining part of the nose section. The team spent most of that day cleaning up the area and digging under the silt for any remaining pieces.

During the recovery effort, they took a piece of the borate tank aluminum back to Fairbanks and flattened it. The team had brought a plaque that had been made back in Michigan and attached it to the piece from the tank, which was then mounted on a piece of well drilling pipe. The plaque was installed and dedicated on June 27 at 2:30 p.m., forty-four years to the minute after the crash. The plaque tells the story of the plane and honors Thorsrud and Gallaher for their careers as aerial fire fighters. After the ceremony, the team went to downtown Fairbanks where the local Experimental Aircraft Association chapter held a banquet and party in their honor.

The following day it was all hands on deck to get the nose section moved from the crash site to the boat. The nose section weighed about a thousand pounds, and it took the entire crew to carry it about a hundred feet at a time, set it down, then carry it again. With the nose section off the sandbar, all that was left was the center section.

The majority of the team departed Fairbanks on July 2, leaving Mihalek, Trainor and his wife, Anna, and Fairbanks locals Fred Aker and Phil Schad. On July 4, the trucking cradle was built for the center section.

On July 5, the helicopter came and in a total of ten minutes the operation was complete. "It flew over, I rigged it, and attached it to the helicopter," Mihalek said. "They picked it up off the sandbar, away it went. Then it went over to Fort Wainwright where we put it on our skid. Todd was at Fort Wainwright with the crew from

Construction Helicopters, and he helped them get it centered as they set it down on the skid." From pickup to delivery at Fort Wainwright, the Sandbar Mitchell was airborne for five minutes, forty seconds. Hemlock Films, producers of *The Restorers*, captured the entire recovery on tape for a future episode.

Josh Palmer, manager of HVAC, Inc., a heating and cooling contractor, donated access to his shop and property for the final preparation of the load before the truck arrived. On July 6, Mihalek and Trainor spent the day removing the leading and trailing edges from the wing so the load would fit within the truck's eight-foot width limit.

On July 7, 2013, everything was loaded on a semi-truck trailer and seven days later it arrived at the museum in Brighton, Michigan.

Moving Forward

Between the remains of the Sandbar Mitchell and the forward and aft section of the sheriff's sale bomber, the Warbirds of Glory Museum has a good amount of the parts needed to assemble a B-25. They will need to acquire engines and propellers, and they are missing one vertical stabilizer.

"We have another horizontal that is in pretty rough shape, but it's from another aircraft that we ended up finding," Mihalek said. "We could possibly rebuild it, but we have a vertical that came off another J model that we got at the Yankee Airforce Museum."

A lot of parts have come from sourcing different places around the country. Todd Trainor was only home from Alaska for a couple of days before he joined another team of volunteers headed off to Colorado to bring back a truckload of B-25 parts from a salvage yard that was going out of business. The Warbirds of Glory Museum bought everything they had. The museum also acquired a fair amount of inventory when the EAA had its warehouse sale, selling off much of its duplicate and triplicate inventory.

"Everything that we've accomplished so far we've done on this 'ramen noodle' budget and it's all been done by people donating stuff.

Recovery of this center section will enable another B-25 to take to the skies. The trip by road took seven days from Fairbanks, Alaska, to the Wings of Glory Museum's headquarters in Brighton, Michigan. *Courtesy Warbirds of Glory Museum*

One of the items we needed was the attach angles that go between the bomb bay and the upper portion of the fuselage and the wing. There was a modification kit developed by the Hayes Co. to strengthen those angles because the North American ones were cracking and they're weak," said Mihalek. "That was one of the things we thought would be very hard to come by. We know we can buy them, but they're expensive. We had a gentleman contact us from our Facebook page whose father worked at the Hayes Co. Sitting in his garage was a brand new upper attach angle kit from Hayes in the original crate with all the North American type orders, bags of rivets, everything. He generously donated it to the museum, and that was one more expensive part we didn't have to buy."

The center section from the sheriff's auction B-25 is too corroded to use in a static or flying restoration, but it has yielded a number of parts that can be used as patterns. It is also a great teaching tool to show volunteers how to drill out and install rivets.

"Our goal is to have the actual center section of Sandbar Mitchell in the shop before the snow falls so we can take the nacelles off it, pull the stress panels, pull the tanks out of it, and take it down to its bare structure. Then we can start going through it with a fine tooth comb to make sure there's no damage in the spars," said Mihalek. "From what we can tell, and from what we've looked at already, we don't think there's going to be anything that's serious in the spars. There's actually one bullet hole in the webbing. We've got to make a patch to repair it or replace it. The spar caps are good as well. Then the team will start rebuilding the bomb bay in the center section. Then we're going to work on the nose section, making repairs where necessary, then mating it to the center section. Once back up on its landing gear, we'll continue to work on restoring the aft portion, outer wing panels, and tail group. We're hoping for a ten to fifteen year restoration to airworthy standards."

The Warbirds of Glory Museum plans to paint the Sandbar Mitchell in the colors of the 340th Bomb Group, which flew from Corsica during World War II. They'll apply Eight-Zulu to the tail and the red cowling markings of the 340th's 488th Bomb Squadron, allowing the B-25 to honor veterans while paying homage to the aircraft's history as a fire bomber. The plane will carry the name *Sandbar Mitchell* as well.

As an interesting aside, the Sandbar Mitchell (44-30733) went down the Kansas City assembly line next to another warbird B-25 known as *Panchito* (44-30734), and it is so rare that two consecutively serial numbered airplanes survive to this day.

Traveling a Unique Path

"It's just been an amazing journey so far and everything here has come in line," said Mihalek. "At one time I was looking at acquiring a P-51H project. It would have been a huge step in my aviation career, but the deal fell through. It was really disappointing, but Todd said something better was going to come along, and he was absolutely right. That's the reason I didn't get the Mustang, because if I had gotten that

The center section of the Sandbar Mitchell will be mated to the forward and aft sections of this B-25 that was rescued from an ignominious life as a decoration at a waterpark. When the restoration is finished, the bomber will be painted in the colors of the 340th Bomb Group, which flew from Corsica during World War II, and will wear the name *Sandbar Mitchell. Courtesy Warbirds of Glory Museum*

project I probably would have never recovered the Sandbar Mitchell. The chance to be able to organize a group of people for the recovery, and to accomplish it successfully in a timely manner, that has never really been done before on a very low budget. We were the underdog, and we succeeded.

"We're just this little group of guys that had information. I had a dream and wanted to do it, and did it."

Australian Outback P-40 over the Wine Country

Chris Prevost in the P-40 and George Perez in his Mustang flying over the Napa and Sonoma Wine Country region in Northern California for the lens of Roger Cain. Prevost operates the Vintage Aircraft Co. at the Schellville Airport in Sonoma, and offers air tours in Stearmans, an AT-6, and in other high-performance warbirds through the North Bay Air Museum. *Roger Cain/Photo plane: Mike DeCastro and Jerry Anderson*

Turning back the calendar more than seventy years to August 1942, Ray Melikian was a young second lieutenant living in a tent near the airstrip at Darwin, Australia. He was part of the USAAF's 7th Fighter Squadron, 49th Fighter Group, and on August 23 he took off in a Curtiss P-40 to intercept an incoming Japanese attack. With three other P-40s, Melikian dove into the on-coming Japanese from 25,000 feet, and was credited with destroying one Mitsubishi A6M Zero fighter.

On March 5, 1943, after the unit had moved to New Guinea, Melikian claimed a Ki-43 Oscar, downed near Malahang. He was promoted to the rank of captain and became the commanding officer of the 7th Fighter Squadron in August 1943. The squadron was based at Tsili-Tsili Airfield, approximately fifty miles due east (inland) of Lae, New Guinea. Melikian was later credited with downing a Mitsubishi G4M Betty bomber.

During the war, Melikian flew 228 combat missions and was credited with three confirmed aerial victories. Melikian's log books also show that he made a number of flights during the war in Curtiss P-40N 42-105306.

In the late 1970s, the hulk of 42-105306 was found at Iron Range, North Queensland, Australia, by Ron Lee and Ian Whitney. It then passed to Charles Darby and Jim Pavitt, then on to Judy Pay. After that, the Curtiss fighter was in storage at Pioneer Aero Restorations in Auckland, New Zealand, for a number of years.

Early 1970s vintage matte photo of P-40N 42-105306 in the brush outside of Iron Range, North Queensland, Australia. The aircraft had been fairly well stripped during the war years and abandoned. *Courtesy Chris Prevost*

The P-40 was put up for auction and Chris Prevost was the high bidder. He then traveled to New Zealand to pack and ship his new project home. The fuselage was pretty far along when he acquired the project in 2000. *Roger Cain*

In 2000, the fighter was put up for auction, and Chris Prevost of Schellville, California, won with a bid of $151,000. The prior owners had spent NZ$1.2 million ($1.45 million) reskinning and restoring the fuselage, but the auction lot did not include an engine, wings, tail feathers, wheels, fairings, or cowlings. All of it would have to be acquired or scratch-built, something that Prevost could easily handle. "When it came on the market it was in *Trade-A-Plane*, and then it went to auction with Pricewaterhouse Coopers. I just never thought, frankly, that I would ever be able to afford to buy any kind of warbird," said Prevost.

Prevost operates the Vintage Aircraft Co. at the Schellville Airport in Sonoma, California, the gateway to the world-renown Napa and Sonoma wine country regions. Since 1975, Prevost has offered air tours over the wine country in his Stearmans or his AT-6, and through an arrangement with the North Bay Air Museum, in a P-51D Mustang.

Of the world's P-40 population, Prevost's N model, 42-105306/N540TP, is one of thirteen N models flying with twenty additional P-40s on civil registers the world over. There is even a radial-engine powered Hawk 75 flying with the Fighter Collection at Duxford, England. In addition, there are a number of P-40 restorations nearing completion, and there may be as many as ten of the type joining the fleet in the next few years.

When Others Want to See You Succeed

Hours after the auction, the seller of the P-40 phoned Prevost and mentioned that he had an engine for the plane under restoration at JRS Engines in Minneapolis, Minnesota, and Prevost could have it for $55,000. Prevost phoned back saying he could scrape up an additional $25,000, but not $55,000. To Prevost's surprise, a deal was made and the engine was added to the package. Prevost flew to New Zealand

P-40 rebuilder Ken Hake provided the jig Chris Prevost used to build a new set of wings for 42-105306. This progress photo was taken in 2004. Prevost started with a few original ribs and built the rest of the wings using blueprints. The main landing gear was installed in 2007. *Roger Cain*

and containerized his new acquisition for the long overseas voyage to the Schellville Airport.

"One of the first things I did after acquiring the P-40 was I went down to California Air Frame in Oakland to see Ted Holgerson," Prevost said. "Ted's a great guy, and everybody in the warbird history thinks that Ted Holgerson is their personal best-kept secret. Ted and his dad have been in the aircraft parts business longer than the rest of us know or have even thought about being in the airplane business.

"I went down there one day and I had brought $4,000 with me to buy a DC-3 propeller that I intended to use on this P-40. Ted showed me this box of DC-3 blades and I said, 'Yeah. I need three of these and a hub.' Ted said, 'Well, those are actually PB4Y-2 Privateer blades.' I asked, 'Is this the same blade?' He said, 'No . . . Well, yeah, it's the same blade but they've got dynamic balancers in them. It's your lucky day. I'll sell you those blades for $100 a piece.' Ted has these moments at times.

"I asked Ted how many he had, and he had ninety. I said I'd take them all!

"Ted asked what I was going to do with all of them, and I told him, 'I've got an idea . . .'

"I came back to the shop and I figured out that those props work on P-40s really well. In fact, they look just like the Curtiss props, except that they're half the money and they probably work better. I ended up trading half of those sets off to P-40 restorers in Australia, New Zealand, and to Ken Hake in Kansas for cowlings, fairings, elevators, wing tips, flaps, whatever I needed.

"That was the best $9,000 I ever spent, and I've still got half the blades. I've got to take my hat off to Ted Holgerson. He's always been just super to deal with. And the generosity of people like Ted make it possible for someone who is not a millionaire to put together a warbird project."

Building the Wings, Gaining the Type Certificate

"I spent a lot of time thinking about this airplane and a lot of energy building it," Prevost said. "I built a set of wings and I only messed up one skin, because I dimpled the wrong side. The wings were completely built from blueprints. I didn't have any examples. I used Ken Hake's wing fixtures, as he's in the process of rebuilding five P-40s.

The P-40's Allison V-1710 engine installation is nearly complete. Chris Prevost was able to acquire the restored engine—for very little money—only hours after he bought the P-40 project. The Allison engine he bought for a song had been freshly overhauled by JRS Engines in Minneapolis, Minnesota. *Roger Cain*

Tom Wilson down in Georgia gave a lot of advice. Tom's really talented and has dedicated his whole life to the Curtiss product. He knows what he's talking about and he helped me out, and vice versa."

To build the wings, Prevost began with a set of ribs and one or two mangled spars. He salvaged the spars, but wasn't happy with them so he built all new spars and replaced about half the ribs he started with. He bought all new extrusions so the wings are entirely brand new. A lot of the parts were acquired from Ken Hake and Tom Wilson. The original match angles for the attach keys were in excellent shape. He also made new landing gear fittings and fuel tanks. And although he did use a lot of original castings and parts, everything in the wings is pretty much brand new.

By April 2008 the P-40N was sitting on its gear nearing its first flight in more than sixty-five years. *Roger Cain*

"The wing set that we started with was a pair of Canadian wings," Prevost said. "I don't know if that means anything, but I'll tell you what a set of Canadian wings means to me—it's going to be a set of wings that were cut off with a torch at the middle gun port on each side. The Canadian farmers did that so they could get the airplanes home. They would just chop the wings off so they could get through a gate, and they also cut the tails off. They'd drag them home and pull out whatever they needed around the farm. All of the nuts and bolts had a use, as did the copper tubing, and for $25 or $50, it was a cheap source of hardware in the post-war period. They'd make wagon wheels out of the main landing gear. They just butchered these airplanes."

One day when Prevost was working on the wings, the OX5 Club was touring the antique aircraft on the airport. One of the tour guests was Boardman C. "BC" Reed who commented, "Those are long-length P-40 wings you're building." Prevost replied, "You're right

The P-40's flight crew includes, front row, Shawn Mulligan, George Perez, Chris Prevost, and Sheryl Carlucci, and, back row, Dave Mace, Denny Jones, and Tom Morris. *Roger Cain*

Ray Melikian flew P-40N 42-105306 on numerous occasions while with the 49th Fighter Group. Melikian flew 228 combat missions and is credited with three aerial victories. He later became commanding officer of the 49th Fighter Group's 7th Fighter Squadron. He retired to the Fresno area in California's Central Valley. Prevost painted Melikian's name and kill markings on the side of the aircraft and arranged for the former pilot to fly in his old aircraft. *Roger Cain*

about that, sir. How would you know?" And Reed replied, "I own the type certificate for it." Reed, who was ninety years old at the time, was in the mood to get rid of things, and Prevost was able to acquire the type certificate from the former USAAF wing commander, who was later the Strategic Air Command base commander at Castle AFB in Central California.

"It took me two years to get the limited type certificate in my name," Prevost said, "because the ACO branch in Burbank, California, saw a note on all the limited type certificates that said, 'No new type certificates will be issued after 1962.' It took me almost two years to argue that meant there will be no new serial numbers, but the certificate is, in fact, valid, and all of the existing serial numbers are eligible to be covered under this certificate.

"That was huge groundlaying work to get that done. They reissued the type certificate data sheet, and if you look at the certificate's data sheet, you'll see that the P-40N is listed as a two-seat airplane.

"About the time I finished with the type certificate, I was talking to a museum director back in New York. I had some Curtiss NC-4 flying boat poster-size, original black and white photographs of the engine nacelles and the cockpit that I sent to the museum.

"They were doing an exposé on that airplane and they were missing all these details. They called me and thanked me for sending it to them. I just sent it to them out of the blue because I figured if I held on to them they'd get wrecked, and while I was talking on the phone, an engineer from Curtiss walked in and donated the service bulletins for the TP-40N field conversion—the whole manual. The museum curator sent it to me as a thank you.

"All the controls on my airplane are put together just like the original TP-40N and thanks to that museum, I now have the whole conversion manual. It's as big as the erection and maintenance manual at 150 pages."

Long Tails, Short Tails, and Test Flights

Curtiss built both long and short tail P-40s. The change was made with the P-40K, block 10. All those that came before, P-40Cs, Ds, Es, Fs, and early Ks, were short tails while the P-40Ms and Ns were long tails. The long tails fly a lot better than the short tails as they are more stable in the yaw axis.

"When I flew my P-40 for the first time, on January 24, 2009, I probably had about ten hours of Mustang time," Prevost said. "The

The front cockpit of the restored P-40 looks like it did when flying in the skies over New Guinea. Chris Prevost restored this aircraft to two-seat TP-40 configuration with a second set of controls in the rear cockpit. *Roger Cain*

Mustang and P-40 are different. It wasn't a difficult transition. The P-40's a pretty honest airplane. I certainly enjoyed the test flight though; it was a kick in the ass after nine years of working on it. But the only difference is that the P-40, even the long-tail P-40, when the tail comes down, the tail comes down. The nose is long and it sits at a steep angle, more so than the T-6 by a bunch. It is probably closer in relation to the Stearman in that it sits prominently nose high. When your tail goes down and you transition to a three point, it's a pretty big transition.

"On take off, you can let the tail fly off. It's easier to kick off, by far, than the Mustang. You have to bring the throttle up slow, otherwise you'll overpower the Mustang's tail. The P-40 doesn't care because it has got a lot of rudder authority.

"The P-40 is not a hard airplane to fly. It really isn't. We have a 2,700-foot by 50-foot runway here at Schellville, and the P-40 gets off in about half the distance when fully loaded. The Mustang is a little bit trickier in that length of runway, but that's a different story . . .

"That said, the systems are a little funky, especially the hydraulic system. You have to know how it works because you have limited options if something goes wrong, and it's not automated at all. It's a lot like a T-6 where you select what you want—gear up, gear down, flaps up, flaps down. Then you pull the trigger and it turns the pump on and makes it happen. But if you run out of hydraulic fluid in the P-40, or you get a line break or something, you're pretty much out of luck. The Mustang's landing gear will freefall if unlocked with no hydraulic pressure, where the P-40's won't. If there's nothing locked and you can't get a thousand pounds to kick it out, you're done and you'll have to belly-land the airplane. You have to really be militant about the maintenance on your landing gear and your hydraulic systems, keeping the fluid clean, keeping the levels checked, making sure your hand pumps are working right.

"Knowing the hydraulic system, combined with keeping the temperatures cool when you're flying it, and everything will go smoothly. You just can't drive a P-40 around the sky with the gear down—too slow and it will overheat. On take off, by the time you get the airplane cleaned up, you'll be approaching your maximum allowable temperatures. P-40 pilots learn to fly the airplane in such a way that the temperatures stay cool. You want to put your landing gear out toward the last part of the approach. You've got to try to time it so your gear's down and locked abeam the end of the runway, instead of on the forty-five degree angle approach to the airport. It doesn't fly very well if the gear and flaps are out. It's really dirty, more so than the Mustang."

Prevost is flying the P-40N about sixty to seventy hours a year.

What's Next?

"So that's where that nine years went, assembling the airplane and getting it flying," Prevost said. "In my opinion it's a really good airplane. It's probably an above-average P-40, but it's not a show airplane. It's just a good, structural, absolutely-true-and-correct-to-the-blueprints, airplane. It doesn't have guns and bullets and ammo cans in it. It's a 5,400-pound airplane and I built it to fly. I had more fun building that airplane. It was a totally compliant airplane project and never gave me any problems."

In early 2013, the P-40 acquired Shark Mouth markings, which really stops traffic on the nearby highway. People see the plane's markings and stop to have their picture taken with the World War II combat veteran fighter. *Roger Cain*

So what does a man who's built a TP-40N do next? He builds another TP-40N, of course. Prevost envisions the airplane will be a tribute to Maj. Gerald R. "Jerry" Johnson, who went on to score twenty aerial victories with the 9th Fighter Squadron, 49th Fighter Group. Johnson flew P-39s and P-40s early in the war with the 54th Fighter Group of the Eleventh Air Force while stationed in Alaska in mid- to late-1942. "It's going to be a tribute to Maj. Jerry Johnson," Prevost said. "Not his airplane. I'm not representing that, but I want to do it as a tribute to him."

The fuselage will be completed in the spring of 2014, and Prevost plans to restore it in a polished or natural aluminum finish. "Yeah, it will be cool. Basically it will be in the same configuration as my airplane. I won't cut any corners, but the second time around there are a few things I'll do differently."

Part Three

Lucky Finds: Rare Warbirds in Unusual Places

It started with the rumor of a Mustang in a garage. Then there was the rumor of two P-38s in a barn. How about a P-63 at an amusement park?

Like the Mustang and the Lightnings, the story of the P-63 is true as well. What's even more impressive is that this hidden warbird was restored to flying condition. The P-63A 42-68864, along with a C-87 Liberator, was donated to Springs Park—an amusement park in Lancaster, South Carolina—in the late 1940s, and kids could climb up a ladder and sit in the cockpit. Springs Park was run by World War I ace and industrialist Elliott White Springs. His family business was Springs Mills, known for Springmaid sheets. When the amusement park closed, warbird collector and aircraft broker Don Whittington from Fort Lauderdale, Florida, acquired the Airacobra hulk. It passed through the Commemorative Air Force and was acquired by Bob Pond. After a lengthy restoration, the P-63 flew again on October 2, 1992, and carried the name *Pretty Polly*. Today the Airacobra is maintained in airworthy condition and is part of the Palm Springs Air Museum's collection.

Then there's the New Zealand collector with an Fb.VI Mosquito, a Mustang, one or two Kittyhawks, and a Hudson in a pole barn. Most of his collection was acquired in the mid- to late-1950s, and over the years he has restored them internally. As the steward of a number of airframes, he takes satisfaction in having preserved so much of aviation's history. Eventually, restoring the aircraft will be up to the next generation. The collector remains a private man, having had so many people trying to tell him what's best for his aircraft.

Talk about a double lucky find. . . . On December 14, 1949, a pair of NACA pilots taxied the second prototype XP-82 (ZXF-82 serial number 44-83887) off the concrete at Cleveland and into the mud. The plane was heavily damaged and was subsequently scrapped. Warbird collector Walter Soplata made the first lucky find when he located the aircraft and bought as much of the XP-82 that he could (one fuselage and other parts) and dragged them home. Soplata stored them until April 2008 when Tom Reilly gets credit for the second lucky find when he struck a deal to acquire the aircraft. The ultra-rare Twin Mustang is now under restoration to fly in Douglas, Georgia. The project posts monthly updates on its website: www.xp-82twinmustangproject.com. *NACA photo*

And the rumor persists that there are anywhere from one to fourteen Mk XXII Spitfires in the desert or scrapyards north of Damascus, Syria. There were attempts to purchase the aircraft in the 1980s, and photos were taken, but the deal did not go through. Today's political situation has probably ended any attempt to recover them for the near future.

These aircraft certainly count as hidden warbirds and when they see the light of day they'll be considered "Lucky Finds."

Return of a Uruguayan Air Force Mustang

The Uruguayan Air Force purchased twenty-five Mustangs from the United States in 1950. By the middle of the decade the propeller-driven fighters were phased out in favor of new T-33s and F-80Cs. The remaining flyable Mustangs were sold to the Bolivians, and two aircraft were preserved with 44-63577 going on display at the Museo Aeronáutico outside of Montevideo. *Courtesy Midwest Aero Restorations*

There was a time in the early 1950s when most of Latin and South America was protected by North American Aviation's "120-day wonder." Mustangs were going south from the United States through various military aid programs. Some were "exported" through cloak and dagger means, while a number were refurbished Cavalier models, and others came from foreign air arms when they upgraded to more modern jet aircraft. In all, more than 180 Mustangs ended up south of the U.S. border. If getting them there was relatively easy, bringing back the survivors twenty or thirty years later was another story.

At the conclusion of World War II, the United States made the decision to equip its allies south of the border with Republic P-47 Thunderbolts as part of the American Republics Project (ARP). ARP was essentially a Lend-Lease follow-on program that gave the Army Air Forces, and later the U.S. Air Force, a legal way to transfer aircraft from its inventory to Latin and South American nations on behalf of the Department of State.

The Uruguayan Air Force (Fuerza Aérea Uruguaya—FAU) was offered Thunderbolts, but elected to receive trainers and transports while they made a decision on what fighter aircraft would be best for its air arm. Eventually, the FAU chose to purchase Mustangs in lieu of receiving Thunderbolts as part of a U.S. government military aid program.

In 1950, thirty-five FAU pilots went north to the United States for training on the F-51. In late November and early December of that year, the FAU took delivery of twenty-five F-51D Mustangs. The Mustangs arrived in Uruguay during the opening months of the

Tyrone Elias of Tulsa, Oklahoma, acquired 44-63577 from Uruguay and had the aircraft shipped, fully assembled, as deck cargo, to the United States. The aircraft entered at the Port of Long Beach, California. *Courtesy Midwest Aero Restorations*

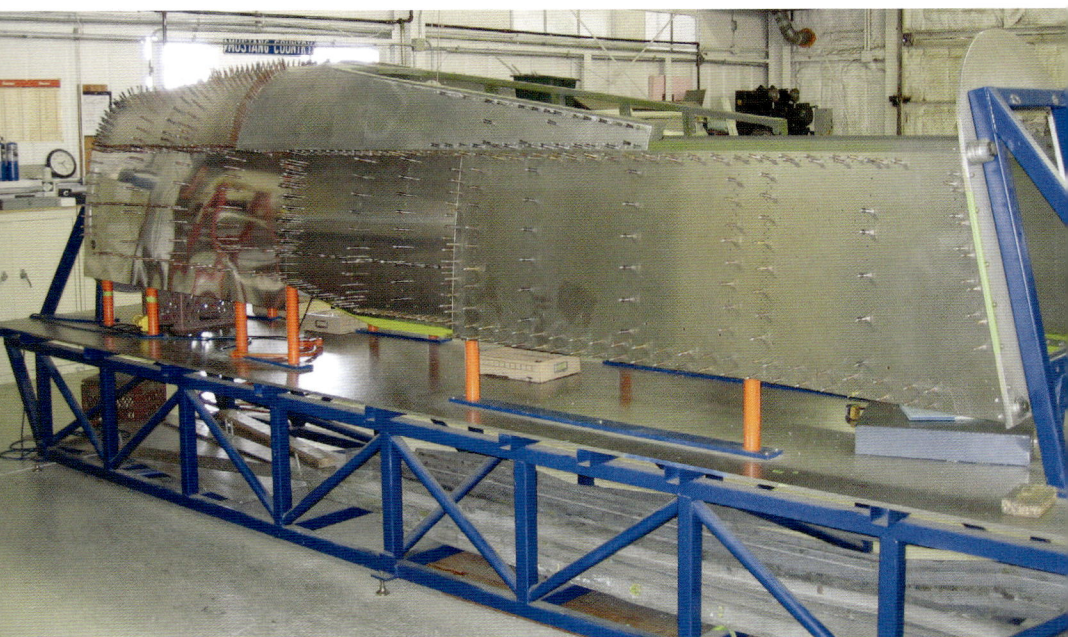

During its service with the FAU, 44-63577 had suffered a number of hard landings and the fuselage had acquired a twist. Once the wing was demated, the fuselage was put into a jig and would not sit correctly. As it was deskinned, the torsion holding the twist released, allowing the structure to line up in the jig. Once the twist was removed, restoration work could begin. Here the fuselage is getting new skin. *Midwest Aero Restorations*

Korean War, and simultaneously the U.S. Air Force had recalled the type for combat duty overseas. Mustang operations in Korea quickly began consuming the U.S. Air Force's stockpile of parts, thus there were none available to transfer to foreign military operators. The resulting parts shortage caused FAU Mustangs to have a very low availability rate.

In 1956, the FAU received its first four Lockheed T-33As, which were supplemented in 1958 with more than a dozen F-80Cs. All of the jets were operated side-by-side with the Mustangs, until the F-51s were eventually phased out in September 1960.

The FAU's five flyable F-51s and associated spare parts were sold to Bolivia. Of the balance of the twenty Mustangs acquired, one went to a technical school and was later sold to a Uruguayan national, three crashed while in FAU service, fourteen were reduced to components or scrapped, and two were preserved in the country.

One of the preserved FAU Mustangs, FAU-270 (F-51D-20-NA 44-63613) served as a gate guard at the Montevideo-Carrasco Airport.

In 1984, it was returned to the United States and is now displayed at the Air Power Museum at Seymour-Johnson AFB, North Carolina.

The second preserved FAU Mustang was accepted on November 20, 1944, at North American Aviation's Inglewood, California-factory, and delivered the same day. From the factory, the Mustang was flown across country to its new assignment, arriving at Sarasota, Florida, on November 28, 1944.

For the next year, 44-63577 flew from training bases in the south-eastern United States—Brookley Field and Birmingham, Alabama, and Jackson, Mississippi. On April 22, 1946, the Mustang was flown to Hobbs Field, New Mexico, for storage with the 4160th Base Unit (Air Technical Service Command—ATSC). The record card shows that on October 28, 1947, the aircraft was transferred to the 4121 Base Unit, Kelly Field, Texas; however, this was probably just a paperwork transfer of "ownership" and the aircraft remained parked at Hobbs.

A little more than a year and a half later, on April 7, 1949, F-51D 44-63577 was transferred to ARP-43 and gained FAU serial number FAU 265, serving until put on display at the Museo Aeronáutico outside of Montevideo.

Odegaard Wings of Kindred, North Dakota, went through the wings, replacing a couple of spars and a number of skins. The wings were shipped back tacked together to enable Midwest Aero Restorations to disassemble and paint each piece before final riveting of the wings. *Midwest Aero Restorations*

Midwest Aero Restorations mechanics—from left, John Chaddock, Gary Day (kneeling), and Dave Young—prepare to mate 44-63577's fuselage to the wings in spring 2013. *Midwest Aero Restorations*

The Long Road to Restoration

In January 1985, FAU 265 was acquired from the Museo Aeronáutico by Dante Heredia of Montevideo, Uruguay. He in turn sold the plane to Tulsa, Oklahoma, businessman Tyrone Elias who had the fighter shipped north, entering the United States at the Port of Long Beach, California.

"When I was attending Spartan School of Aeronautics in Tulsa, Oklahoma, back in 1984 or 1985, I had heard about a local guy who was going to be bringing a Mustang into the country. That was right about the time I was graduating from Spartan and returning to Illinois," said Mike Vadeboncoeur of Midwest Aero Restorations Ltd. in Danville, Illinois. "He did import that Mustang, but I never saw it until I was friends with Butch Schroeder. One day we jumped in Schroeder's gray Mustang (45-11559, *North American Maid*) and flew down to visit Elias because he'd asked for some advice on his Mustang. At the time, Schroeder was involved in rebuilding the photo-reconnaissance F-6 Mustang *Lil' Margaret*, which went on to take a number of top honors. That's the first time I saw Elias's airplane and it was a mess.

"We told Elias that maybe he needed to take the Mustang to a shop that had fixtures and jigs to make the airplane right. At that time, Art Teeters (Cal-Pacific Airmotive, Salinas, California) was the man who was doing a lot of the major, deep structural work on Mustangs. Elias ended up not doing that. That was the first time I saw the airplane and then, of course, it ended up in our shop all these years later, which is kind of cool."

Elias sent the Mustang to Nelson Ezell's shop in Breckenridge, Texas, where they did some work on the airplane. Shortly after leaving Ezell's shop, the Mustang was sold to the Whittington brothers who purchased it to resell it.

"John Turgyan then bought the Mustang (in 1994) and that's when I became reacquainted with it again. I met him while working for Butch Schroeder," said Vadeboncoeur. "Turgyan had all the intentions in the world of putting the airplane back together. It was a slow process as he was collecting parts and trying to have pieces rebuilt and pieces repaired that maybe weren't repaired very well in the beginning. About 2008, he decided to send the aircraft to me for a ground-up restoration. We got it in the door and had started working on the plane when Turgyan sold it in 2012. That's when the current owner, Jon Vessly, stepped in and he decided to finish the project with us.

"When we got started on the ex-Uruguayan Mustang we suspected the fuselage had a pretty significant twist in it along with some other repairs that were not quite what they needed to be, so we basically had to start over with a lot of things. We put the fuselage into our fixture and it wouldn't line up, so we started unriveting the skins. After we took the skins off of, which was, of course, holding in the twist, the fuselage just fell right back into place like it wanted to be there. Once we deskinned about ninety percent of the fuselage, it just settled right in. Then we stripped it. During the previous restoration it had been painted all gray, which was the thing to do back in the 1980s. John Vesely wanted it more authentic, like what we did with *Happy Jacks Go Buggy* (44-74452, N74190), so that was our direction from him, and of course that meant going backwards in a lot of areas.

"The wings had some other significant issues and we ended up sending those to Brent Meester at Odegaard Wings (Kindred, North Dakota), and they did some spar replacements and replaced certain skins. We have Meester's crew take the rebuild to a certain point, then they send the assembly back to us for completion. For instance, in

Sitting on jacks, 44-63577 comes together. This was Midwest Aero Restoration's first aircraft painted at their shop. The fuselage and wings were painted separately, and the wing fillets and cowling were fitted soon after this photo was taken, then painted. Close attention to detail ensured a perfect color match on all of the assemblies. *Midwest Aero Restorations*

the wheel well, we didn't have them rivet any of the parts together, we just had them fit everything while it was in their fixture. When it gets back to our shop, we paint all of the parts separately, then shoot them together. It gives the aircraft a factory-finished look." Midwest Aero sent the engine mount to Meister as well.

In the past, Midwest Aero Restorations's Mustangs have been finished in natural metal or polished metal with markings, but 44-63577 was the first aircraft the shop has completely painted. "We decided to go ahead and roll the dice and hope that everything would match up, and our paint process is such that we keep a pretty careful eye of how we apply the paint, the conditions that we apply it in, and how we mix everything," said Vadeboncoeur. "I think it matched out really well. You always run a risk, especially with metallics, that you get a different look from one day to the next or one part to the next. So, yeah, we painted the fuselage separate, we painted the wings separate, and then some of the cowling and fairings, which we can't paint until after the wing and fuselage are mated. Once the fuselage and wings are mated we can fit the cowling and fairings, then take them off and paint them. We were kind of concerned that we were going to have match problems, but the process we are using worked out just fine and it all blended really nice."

When John Turgyan owned the aircraft, he spent a lot of time and money hunting down original Mustang parts to give the restoration a super-authentic look. Turgyan had planned to restore the aircraft to resemble Capt. Arval J. Roberson's 357th Fighter Group P-51D *Passion Wagon*. Turgyan had known Roberson, and out of respect to Turgyan—should he ever decide to pursue another Mustang restoration—Vesely and Vadeboncoeur began looking for a new paint scheme for the ex-Uruguayan airframe. "I kept coming back to something out of the Korean War era. When you Google 'Korean War Mustangs,' this is one that pops up pretty frequently, in the modeler world. In all that research we saw this airplane in color and we narrowed it down between this and another Mustang. We liked this one because it had the shark mouth with the eye and it had a name, *Was That Too Fast*,

Building a Restoration Reputation

If Mike Vadeboncoeur's name or his business, Midwest Aero Restorations, sound familiar, it's because his shop and his staff have built a number of award-winning restorations during the past two dozen years. Vadeboncoeur won the Experimental Aircraft Association's coveted Golden Wrench Award at the EAA AirVenture Fly-in in 1990 for his work on Butch Schroeder's Grand Champion AT-6.

In 1993, Vadeboncoeur and Midwest Aero Restorations began a streak of awards for P-51 restorations starting with the restoration of Butch Schroeder's F-6D *Lil' Margaret* (44-84786); for Ken Wagnon's 2002 Grand Champion World War II Warbird winning P-51D *Cripes A Mighty IV* (44-74813); for Charles Somers's 2004 Reserve Grand Champion *Daddy's Girl* (44-63807); for Dan Baun's 2006 Reserve Grand Champion *Red Dog* (44-74469); and Bruce Winter's 2008 Grand Champion World War II Warbird and Rolls-Royce Aviation Heritage Trophy winner *Happy Jack's Go Buggy* (44-74452).

Phoenix businessman Jon Vesely owns P-51D-25-NA 44-73343, that was restored by Midwest Aero and completed in 2010. At that year's EAA AirVenture Fly-in Vesely's Mustang was awarded Reserve Grand Champion World War II and Midwest Aero Restorations earned another Golden Wrench Award. Vesely's award-winning D-model wears the nose art of *Live Bait*.

Suffice it to say that Midwest Aero's P-51 restorations are helping to set the standards for authentic Mustang rebuilds.

which is appropriate for a Mustang. It also had the squadron emblems that were painted on the vertical and on the side of the fuselage, which was unique to this airplane. It had a lot going on and we thought that would be really attractive," said Vadeboncoeur.

The original *Was That Too Fast* served with the 18th Fighter-Bomber Wing (FBW), 12th Fighter-Bomber Squadron (FBS), which flew F-51s from Chinhae Air Base, known as K-10, South Korea. K-10 is on the southern coast near Pusan, and Mustangs from the 18th FBW would stage through K-16 (Seoul) and K-55 (Osan) to attack targets in the north. The 12th FBS Mustangs were called the *Foxy Few*, and their unit emblem was painted on both sides of the fuselage and the tail.

Opposite and above: When John Turgyan bought the aircraft in 1994, his vision was to bring the Mustang back as close to factory stock as possible. Many of the parts he acquired were used by Midwest Aero Restorations to detail the cockpit. Note the rocket control box to the left of the stick and mounted below the main panel under the clock. *All Midwest Aero Restorations*

"When we decided to go with a Korean War Mustang, I immediately thought about ordnance and how cool it would be to have some bombs and rockets on it," said Vadeboncoeur. "If you notice, the rocket rails that are on the airplane are not what you would typically see on a P-51. Up through the P-51D-20 they did not have provisions from the factory for rocket launchers. Dash 25 airplanes and later, did. The ones where you see the individual stubs bolted to the wings, those were all Dash 25 airplanes or later. Operations in the Korean War called for aircraft to deliver rockets against ground troops and other targets. For all the airplanes that didn't come out of the factory with rocket launcher provisions, they developed a field modification kit. Jay Wisler, of Warbird-parts.com in Tampa, Florida, had tried to sell me one of those kits for years and I always ignored it because you never see them on World War II airplanes.

"This is a Dash 20 airplane that didn't have any provisions for rockets, so I bought the kit from Wisler, opened it up, and it was brand spanking new, original—every last nut, rivet, bolt, wire, bracket, you name it, to bolt these things to the bottom of the wing that wasn't equipped for them. The rocket launching kit mounts to the underside of the wing with a one-eighth-inch plate. The load is distributed across the wing, and using a mylar template included in the kit, you drill out the existing rivets and reach into the wing and install nut plates or nuts to secure the kit into existing ribs and stringers. It was really neat to be able to bolt the kit on the airplane and hang ordnance from it."

For the restoration, Vadeboncoeur's crew worked to keep the cockpit completely stock. The rocket control kit selector control panel was added, a fuel flow meter was added for safety, and basic modern avionics were added under a panel used to control the ARC-3 radio. The ARC-3 radio can be seen sticking up into the cockpit behind the pilot's seat, and this model replaced the World War II-vintage SCR-522 set.

Was That Too Fast is powered by a V-1650-9A Merlin engine built up by Mike Nixon's Vintage V-12s in Tehachapi, California. The engine was restored a number of years ago and had sat for a long time. It was sent back to Vintage V-12s, where it was opened up, inspected, reassembled, and the engine's exterior color changed from gray to black to more accurately represent a Korean War-era Mustang.

Now registered N151JT, the Mustang made its first post-restoration flight on July 11, 2013, with well-known Mustang pilot Vlado Lenoch at the controls. Lenoch's flight in 44-63577 was the aircraft's first flight since the early 1950s as the Mustang spent most of its time in Uruguay sitting on display. Using flight hour data from a group of Mustangs imported from Guatemala, Vadeboncoeur provided an estimate that 44-63577 has only 1,300 hours on the airframe, and when one considers that the FAU only flew the type for a decade, this plane much less, in actuality it could have far less time in the air. For Federal Aviation Administration regulations, 1,300 is the aircraft's official airframe total time and was entered into the fighter's log books after the restoration.

Following its initial flights, Vadeboncoeur's team detailed the fighter and put time on it in preparation for its debut at EAA AirVenture 2013. AirVenture was held July 29 to August 4, and when *Was That Too Fast* arrived, there was a lot of buzz about the new

To complete the aircraft's authentic look, Midwest Aero Restorations located a new old stock rocket retrofit kit for the F-51D-20 and earlier model Mustangs. The retrofit kit can be seen in this view—a one-eighth-inch plate mounts to the underside of each wing, outboard of the bomb hard point. Rivets in the wing are removed and replaced with bolts that mount into the wing's existing ribs. The zero-length rocket rails then mount to the plate. *David Leininger/Photo plane pilot: Mike Shore*

Mustang. EAA's Warbirds of America hosts one of the most prestigious warbird competitions, and the judging is of the highest caliber. Aircraft are judged on eight categories (General Appearance; Fuselage; Landing Gear and Wheel Wells; Wings and Tail Surfaces; Engine and Accessory Section; Cockpit; Authenticity; Depth of Restoration) with a maximum of one hundred points to be awarded for these areas (they do have a tie-breaking points scheme as well should two or more aircraft achieve the same scores). When the judging was completed and the scores were tallied, Jon Vesely's Korean War F-51D *Was That Too Fast* was honored with the Grand Champion Post-World War II Warbird award and Mike Vadeboncoeur's Midwest Aero Restorations was recognized for its work with the Golden Wrench Award.

After *Was That Too Fast* was moved out of the shop, Vadeboncoeur and Midwest Aero began working on a couple of warbird repair projects and have another ground-up Mustang restoration in the wings. During the next five years, they'll complete the restoration of one of the Mustang's nemesis, a World War II combat-veteran Messerschmitt Bf-109.

Three months after winning the EAA Warbirds Grand Champion-Post World War II award at AirVenture 2013, Vesely put *Was That Too Fast* up for sale, asking $3 million. Whoever buys this aircraft will probably consider it cheap at that price, and they'll be thinking they made a lucky find, too.

Hawker Tempest II
Restoration in Texas

This factory-fresh Hawker Tempest II, RAF serial MW404, is seen in July 1945 at Hawker Aircraft Limited's Langley, England, factory. The aircraft subsequently served with the Royal Indian Air Force as serial number HA557. The Indians flew Tempests until 1953, when they were phased out in favor of de Havilland Vampire jet fighters. *Courtesy Chris Miller*

Invasion! Two of the world's population of eleven surviving Hawker Tempests have been imported to Texas and a third is on display in Florida. That's a British invasion . . . of sorts . . . when you consider that those three aircraft make up twenty-five percent of the survivors of that type.

Kermit Weeks owns a Tempest II that is on display at the Florida Air Museum at Sun 'n Fun in Lakeland, Florida, and Nelson Ezell is rebuilding a Tempest II in Breckenridge, Texas. The third was recently imported by Chris Miller, and it is now undergoing restoration at the David Wayne Hooks Memorial Airport, north of Houston in Spring, Texas. Miller is a civil and mechanical engineer during the day and a warbird pilot and restorer on the side. He most recently was working on an AT-6 project and is now fully occupied with rebuilding Hawker Tempest II, RAF serial MW404.

Tempests IIs are large, single-engine fighters that were an outgrowth of the Hawker Typhoon and Tempest Mk V. External differences between the Typhoon and the similarly looking Tempest Mk V show that the Tempest II has a longer fuselage (34 feet, 5 inches for the Tempest II and 33 feet, 8 inches for the Mk V), a thinner, elliptical, laminar flow wing, and the Mk V's chin radiators were relocated to the wing leading edges on the Tempest II. The Tempest II was fitted with four, wing-mounted 20mm cannon and could also carry two thousand pounds of ordnance.

The prototype Tempest II first flew on June 28, 1943, and it was so promising that the RAF ordered 2,630 of the type to be built by Hawker (1,800), Gloster (500), and Bristol (330). However, production was delayed when the decision was made to fit the aircraft for service in the tropics. The type was slated to see service in the Pacific against the Japanese as part of the British Commonwealth's "Tiger Force," a very long-range bomber group that was to be based on Okinawa using Lancasters, Lincolns, and Liberators. The Tempest IIs would escort the bombers to and from targets in the Japanese home islands and other occupied territories. The atomic bombs dropped on Japan ended the war before Tiger Force could be assembled and the Mk II could enter combat.

Previous versions of the Tempest, in particular the Napier Sabre-powered Tempest Mk V, had excellent low-altitude performance. They served as ground-attack aircraft and were used to run down German V-1 buzz bombs. The Mk II is powered by a 2,590-horsepower, 18-cylinder sleeve valve Bristol Centaurus V engine turning a four-blade, 12-foot, 9-inch propeller, which gave the aircraft a top speed of 442 mph at 15,200 feet. The Centaurus engine proved to be much more reliable than the Napier Sabre under the demands of flying combat operations, thus the Mk II's success. In total, 452 Tempest Mk IIs were built—136 standard aircraft and 316 Fighter Bomber configured Tempest IIs (designated as FB.II). The follow-on Hawker Sea Fury is essentially a lightweight version of the Tempest II with smaller, folding wings (38 feet versus 41 feet for the Tempest II), a raised cockpit for better forward visibility, and the Sea Fury's maximum gross takeoff weight (12,498 pounds) is 1,402 pounds less than the Tempest II (13,900 pounds). The two aircraft do, however, share many of the same components and sub-assemblies.

Tempests in the Far East

In 1947, India was to gain its independence from Britain, and although the Royal Air Force would remain in the country, the Royal Indian Air Force (RIAF) would need its own modern equipment to maintain its sovereignty. In the lead-up to independence, the RAF was converting a number of its squadrons based in India from the Spitfire to the Tempest II. RAF No. 5 Squadron was the first to transition, followed by Squadrons 20, 30, and 152 converting to Tempest IIs in June and July 1946. At the end of the year, these squadrons gave their aircraft over to the RIAF and subsequently disbanded. In total, the IAF received eighty-nine Tempest IIs.

To maintain a balance of power in the region, the Pakistani Air Force (PAF) asked the British for twenty-four Tempest IIs. PAF squadrons 5, 9, and 14 received a total of twenty-four Tempest IIs, which were based at Lahore, Miranshah, and Peshawar.

Although the type did not see service in World War II, they were extremely active in a number of small, border conflicts. RAF Tempest IIs of No. 33 Squadron flew ground attack missions against terrorist groups in the jungles of Malaya in late 1949 and 1950.

The RIAF took its Tempest IIs to battle for the first time in October 1947, against insurgents near Badgam, in the Kashmir region of India. The Tempests of No. 7 Squadron (RIAF) strafed insurgent positions and the truck convoys that supplied them. In October 1948, Tempests prevented Indian Army troops from being overrun at Tithwal by attacking enemy troops, gun, and mortar positions with cannon and rocket fire.

The Tempests of No. 3 Squadron and No. 4 Squadron also supported the army during Operation Polo, in the Hyderabad state in July 1948. Tempests bombed an insurgent airstrip and took out numerous enemy strongholds. On January 26, 1950, the RIAF became the Indian Air Force.

The Indians lost twenty-seven Tempest IIs between August 28, 1947 and April 28, 1954. Some were training losses while at least eight were lost during operational sorties flying against insurgent forces during the Kashmir Conflict. Tempests were flown until 1955 when the aircraft of No. 4 Squadron, the last IAF squadron to fly the fighter-bomber, were phased out in favor of the de Havilland Vampire F.B.52 jet fighter.

Eleven Hawker Tempest IIs that had served as decoys were offered for sale by the Indian Air Force in 1975. The majority of the aircraft had the wings demated from the fuselages and they were heavily over-grown. At the time of the sale, the four Spitfires offered attracted all of the attention and warbird collector Doug Arnold bought the ignored Tempests. Arnold selected the best of the lot and had the remaining aircraft scrapped to keep a tight hold on the Tempest market. *Peter Arnold*

Bringing Home the Tempest IIs

In September 1975, the Under Secretary to the Government of India circulated an offer for a sealed-bid sale of "obsolescent/obsolete aircrafts and allied equipment." The list contained fifty-six Harvards, thirty-five Mystère IVAs, forty-two Vampire F.B.52s, thirty Vampire T55s, five Consolidated Liberators, four Spitfires, thirty-eight Ouragans, and eleven Tempest IIs, plus a large number of engines, some counted as still being in the aircraft—248 Rolls Royce Goblin engines (for the Vampires), 34 Verdun engines (for the Mystère), 116 Nene engines (spares for the Ouragans), and 79 Pratt & Whitney R-1340 engines for the AT-6/Harvards. Spare parts were also included, but they were not called out separately in the sale offering. Bidding was scheduled to close on February 29, 1976.

When these lots were offered, their value was essentially that of a huge pile of scrap metal. The four Spitfires held the most interest,

Chris Miller recently imported Tempest II MW404 to the Houston, Texas area. Miller's acquisition adds a third of the high-performance fighters to the U.S. warbird community—Kermit Weeks and Nelson Ezell owning the other two. The fuselage of MW404 will be split at the production breaks into three sections, restored separately, then reassembled. *Chris Miller*

while everything else was considered "run of the mill." Reportedly, the engines from all of the AT-6/Harvards were removed and sent to the United States, while the fuselages and the spare parts were simply abandoned in place. Keep in mind, at this point in time, the AT-6/Harvard was still in fairly wide-spread service with the Portuguese and South African Air Forces, among others.

The Tempests were hidden in the brush outside the airbase at Pune (previously called Poona), with their wings removed, sitting on their bellies. At the time, a Tempest II project was valued at around $25,000 each, if a buyer could be found. In 1979, warbird collector Doug Arnold acquired seven of the best IAF examples and had the remainder scrapped or bulldozed to ensure the remaining four would not come on the market. Not all of the wings matched the serial numbers of the fuselages, as Arnold cherry-picked what he could, so the identity of the surviving aircraft corresponds to the fuselage serial number. Arnold imported his cache of Tempests to England, and then set about trying to turn a profit on his investment.

Arnold also picked up one of the B-24 Liberators (B-24J-85-CF 44-44052) from the sale list, which had been used in a technical school. It was subsequently sold to the Collings Foundation in 1984, then completely restored, and today flies as *Witchcraft*.

Arnold successfully sold off his Tempest II collection with many going to warbird collector Nick Grace and Chris Horsley's Tangmere Flight, based at Tangmere, England. Tangmere Flight traded one aircraft (Tempest II PR536/IAF HA457) to the Royal Air Force Museum for a Napier Sabre engine and Tempest V fuselage (RAF serial EJ693) that the museum had acquired from Holland. When Nick Grace was killed in an automobile accident, Horsley retained a number of the aircraft.

Making the Deal

"I had been looking for a warbird project for probably a good year, and there was a P-39 project for sale," said Chris Miller. "I nearly bought that one, but that deal did not go through. Then I just started looking more and more, and looked at a Yak-3 project and I wasn't happy with what I saw. It was the only thing I could find at the time. Then I just told myself to wait a little longer and I just kept looking.

"I happened to be selling my T-6 project the same time. It was about ninety percent finished when I ran into a friend who mentioned that there was a Hawker Tempest for sale in England and it was owned by the Horsley brothers. I started looking for it and eventually found it. I went there late in 2011, and it looked really interesting. It took about six months of negotiating to make the deal."

The seller agreed to crate the plane, and as he was loading it, it quickly became clear that the wings were too wide and would not fit into a normal shipping container. Each wing weighed about a thousand pounds so repositioning them each time was a chore. The close-topped container was swapped for a open top unit and the wings were loaded at an angle, which solved the problem.

When the containers arrived in Texas, what Miller found was a Hawker Tempest II, RAF serial MW404 that was built in July 1945 at Hawker Aircraft Limited's Langley, England, factory. In August 1945, the fighter was delivered to No. 247 Squadron at Chilbolton. It flew

The wings look in pretty good condition considering they were exposed to the weather for more than thirty years. Notice the wing's cannon shell ejection ports at the top left of the photo. *Chris Miller*

at Chilbolton until transferred to No. 20 Maintenance Unit in spring 1948. In May of that year, it was acquired by Hawker Aircraft and refurbished for service in the tropics with the Indian Air Force. The IAF re-serialed the fighter HA557, and it flew until retired in 1953. From then until recovery by Doug Arnold, HA557 sat outside at Poona, some say as an airfield decoy, but that job did not last long as the fighter was soon covered with plant life. James and Chris Horsley acquired the fighter in 1988, and it was stored while they worked on other projects. Miller acquired the Tempest in 2012.

Miller has broken the fuselage down into components, and the steel structure, much like a Hawker Hurricane from the tail cone forward, is in three sections. The plan is to restore one section at a time. Miller knew there was corrosion in the lower fuselage, and he estimates that he'll have to replace one-quarter of the chromoly steel tubing. Both wings and the rear fuselage are in now in jigs, and reskinning of the rear fuselage has been started. There has been a lot of trial and error involved due to very thin skins and light construction.

Working in parallel, Miller is also making templates for the rudder ribs. He located a rudder post that has the leading edges for the ribs, which are intact. The rear section of the ribs are simple triangles,

Detail photo of the Tempest's vertical stabilizer and tail wheel assembly. Miller will have to scratch-build the horizontal stabilizers, elevators, and rudder as they were either not recovered from India or were lost during years of storage. *Chris Miller*

The wings are in good condition, although they are not the set that were originally installed on MW404/HA557. This set is from HA407. When the aircraft were recovered from India, the best fuselages and best wings were brought out, and the rest scrapped. *Chris Miller*

and those will be stamped out using the new templates. Tempest owner Nelson Ezell has given Miller a pile of surplus ribs to be used as patterns for the horizontal stabilizers.

Miller plans to forego installing the complex and difficult-to-maintain Bristol Centaurus engine with a Wright R-3350 radial and a propeller from a Skyraider. He traded his nearly complete AT-6 project for the powerplant and prop.

"The wings that came with my fuselage (HA557) are not from my aircraft," said Miller, referring to how the components were mixed when they were recovered from India. "The wings are from one of the fuselages that was scrapped (RAF PR555/IAF HA407), and they appear to have some battle damage repairs. They look like they were hit by a .30-caliber machine gun bullet that went into the oil core inlet and disintegrated, hitting the main wing spar. The spar has some little dents in it, which are repairable. There is also a .30-caliber bullet patch in the other wing near the aileron linkage."

In total, there are eleven surviving Tempests, nine Mark IIs and two Mk Vs. Of the Mk Vs, RAF serial EJ693 is owned by Kermit Weeks's Fantasy of Flight Museum in Polk City, Florida. The fighter-bomber is under restoration to airworthy standards with Personal Plane Services in England. Weeks is having it finished as a static display for now, and will come back and complete the systems work to make it a flyer in due time. The Mk V was powered by a twenty-four-cylinder, sleeve valve, 2,240-horsepower Napier Sabre IV engine where the cylinders are set in an inline "H" rather than a "V" or radial configuration. Weeks has the only two Napier Sabre engines in public hands, so if anyone can make it fly, he certainly has all of the parts necessary to make it happen. The other Tempest Mk V is displayed at the RAF Museum in Hendon, England, suspended from the ceiling and wearing target tug markings.

Aside from the three Tempest IIs in the United States, MW848/HA623 is preserved at the Indian Air Force Museum at Palam, New Delhi, and the RAF Museum at Hendon showcases PR536/HA457. Tempest Two Ltd. owns three others, and MW376/HA604 had been under restoration with Pioneer Aero Services in New Zealand, until the aircraft's owners perished in a flying accident.

Miller has a long restoration ahead, but who knows, his maybe the first of the Tempests to fly. It will take a few years before a Tempest II gets airborne, but it will be a phenomenal sight when they take to the air.

Hawker Tempest Survivors

RAF Serial	IAF Serial	Mark	Registration	Owner/Location	Status
LA607		II	N607LA	Kermit Weeks/Displayed Florida Air Museum at Sun 'n Fun, Lakeland, Fla.	PV
MW376	HA564	II	G-BSHW	Estate of Eric Hertz/under restoration at Pioneer Aero Services, New Zealand	S
MW401	HA604	II	G-PEST	Tempest Two Ltd., England	S
MW404	HA557	II		Chris Miller, Houston, Texas	R
MW758	HA580	II		Tempest Two Ltd., England	S
MW763	HA586	II	G-TEMT	Tempest Two Ltd., England	R
MW810	HA591	II		Nelson Ezell, Ezell Aviation, Breckenridge, Texas	R
MW848	HA623	II		Indian Air Force Museum, Palam Air Base, New Delhi	PV
PR536	HA457	V		Royal Air Force Museum, Hendon, England	PV
EJ693		V	N7027E	Kermit Weeks (under restoration at Personal Plane Services, England)	R
NV778		V		Royal Air Force Museum, Hendon, England (in target tug markings)	PV

Sources: Christer Landberg, www.hawkertempest.se; Geoff Goodall, Warbirds Directory; plus additions by the author.

F = Flyable; PV = available for Public Viewing; R = under Restoration; S = in Storage

Where Did All the B-24 Liberators Go?

Missing B-24s from Schools and Memorials and the South American Connection

Armonk Air Squadron One, from Armonk, New York, received this ex-U.S. Navy RY-1, BuNo 67797. Scoutmaster K. V. Tindall and six other Scout Council members chipped in and raised the $300 necessary to buy the former bomber-transport. The aircraft is seen shortly after its December 1946 arrival. It became a clubhouse for the scouts. *William T. Larkins Collection*

In the months following the end of World War II, thousands of surplus warbirds sat at depots across the United States. Before America's surplus aerial armada* was scrapped, fighters, bombers, and their components were offered to schools and civic groups to serve as memorials and teaching tools.

When it came time for cities, towns, and schools to select a four-engine bomber for their technical schools or as a war memorial, most turned to the Flying Fortress. Today, twenty-one B-17s have been tracked from post-war storage yards to schools and towns across the United States.

The Boeing B-17 Flying Fortress had been given a great deal of exposure as a USAAF propaganda tool, and the aircraft's exploits were illuminated on movie screens across the country during the course of the war. Saturday afternoon matinee newsreels and feature films showed the B-17F *Memphis Belle* attacking targets in Nazi Germany, while dashing actor Clark Gable served on board the B-17s of the 351st Bomb Group at RAF Polebrook. On the silver screen, actors John Garfield, John Ridgely, and Gig Young flew a B-17C named *Mary Ann* into the attack on Pearl Harbor, and then took the fight back to the Japanese in the movie *Air Force*.

The Consolidated B-24 Liberator, often derisively called "the box the B-17 came in," had an outstanding combat record during World War II, and had served with distinction in all theaters of war. It was, however, pretty much ignored by the American military's publicity machine, and after the war was quickly phased out of service by the USAAF in favor of the forthcoming jet-powered bombers. The U.S. Navy, on the other hand, kept its twin-tailed PB4Y-1 version of the B-24 Liberator in service until the mid-1950s. Without much glory and still seen as a flying workhorse in post-war service, there was little interest in honoring the Liberator. It was considered old by

B-24J 44-40238 *Thunderbird* saw combat with the 852nd Bomb Squadron, 491st Bomb Group at North Pickenham, England, and was sent to storage at Altus AAF, Oklahoma, after the war. From Altus, it was flown to the Flandreau Airport, Flandreau, South Dakota, where it was donated to the Civil Air Patrol squadron. It was later inherited by the Boy Scouts. Derelict and neglected, it was sold for scrap in 1957. *Al Lloyd Collection*

A young Larry Kotz and his family toured B-24J 44-48834, which had become a war memorial for the Edsel B. Ford American Legion Post. The aircraft suffered heavily from vandals and the weather, however at the time this photo was taken the top turret dome was intact. The fabric-covered rudders were also gone by this time. When the VFW Post closed in 1950, the aircraft was reportedly scrapped. *Larry Kotz*

technological standards thus few aviation trade schools chose to use them as instructional airframes.

The U.S. government assigned the task of disposing surplus aircraft to the War Assets Administration, which later became the Reconstruction Finance Corp. In May 1945, regulations set forth the costs and the procedures that enabled surplus war materiel to be transferred to schools and civic groups at minimal expense. For example, these groups could acquire a B-17 or B-24 for $350 plus delivery costs; a medium bomber type, like a B-25 for $300; and fighters like a P-51 Mustang for $500. The only restriction was that the U.S. government still retained ownership of the aircraft and the plane could not be flown. When the municipality or school was done with the aircraft, the government required that it be scrapped.

In all, at least ten Consolidated B-24 Liberators were acquired from surplus stocks to serve as memorials, instructional airframes, and for use by civic groups. Eight of the ten have been identified, and while photos of the others have surfaced, their identity is still a mystery. Where did these aircraft end up? If they were scrapped, surely there must be pieces still left? And if the bombers could not be flown, how did so many end up in Latin and South America? Another aircraft, C-87 42-107257 was bought surplus and ended up as the snack bar at an amusement park in Lancaster County, South Carolina. More than seventy years after the beginning of World War II, many of the surviving Liberators owe their existence to the schools, memorials, and south of the border operators that found a use for surplus B-24s.

The Boy Scout Bombers

In the United States, the Boy Scouts are a rite of passage for many young men. Founded in February 1910, the Boy Scouts teach civic, social, and wilderness skills to young men throughout the world. In post-World War II America, service to country was a big part of scouting, and after scouting many young men went on to join the

Reserve Officers Training Corps in high school or college. Three scout troops acquired Liberator bombers in late 1946 and early 1947 for the dual use of war memorial and clubhouse.

The Boy Scout troop in Urbana, Ohio, was one of the first to receive a B-24 Liberator through the Reconstruction Finance Corporation's educational and memorial aircraft program. Ford Willow Run-built B-24J 42-50801 was delivered to the Army Air Forces on May 31, 1944. The aircraft remained stateside flying from Brooklyn, New York; Olmsted Field, Pennsylvania, and finally with the 4000th Base Unit—a replacement aircraft pool, at Patterson Field, Ohio. This Liberator was delivered directly from Patterson to Urbana on September 27, 1946, under the auspices of the Reconstruction Finance Corp.

Near the border of New York and Connecticut lies the hamlet of Armonk, New York. In the summer of 1946, Armonk Scoutmaster K.V. Tindall and six other men contributed a total of $300 to acquire a surplus B-24 for use as a clubhouse. The aircraft delivered to the young men of Armonk Air Squadron 1 was former U.S. Navy RY-1 BuNo 67797 (ex-C-87A-CF 43-30569). The aircraft was flown from Bush Field, Georgia, to the airport in nearby Purchase, New York, and then towed by road to the small airport at Armonk. This aircraft was fitted with a sixteen-passenger interior, one of only three such conversions acquired by the Navy, and was ideal for a clubhouse. Years later, when the plane's condition had deteriorated, it was scrapped.

The final Boy Scout bomber is somewhat of a mystery. In late 1947, local businessman Allen H. Sims presented the bomber to the Piedmont Council of the Boy Scouts for its troop in the Gastonia, North Carolina, area. He had gathered $350 and presented the aircraft in memory of his son. Through photos and the memories of a number of Gastonia-area Boy Scouts, the identity of this aircraft has come down to a few possibilities: former scout Paul Quinn remembers the aircraft having a metaled-over nose, much like a C-109, and he's seen a photo of TB-24M 44-42535 at Hagerstown that might be the Gastonia bomber.

Other possibilities for the Gastonia Boy Scout bomber are B-24M 44-51901 and the combat veteran Ford-Willow Run-built B-24M 44-50960. B-24M '960 was delivered to the USAAF on February 19, 1945, and underwent modification at St. Paul, Minnesota. On April 19, the Liberator departed Mather Field, outside Sacramento, California, for the Pacific Theater. Christened *Tricky Mickey*, this B-24 flew with the Fifth Air Force's 90th Bomb Group, and later with the 7th

B-24M 44-51556 had served for four months in the Panama Canal Zone before returning to the United States, to Kingman, Arizona, for storage. From Kingman, the Liberator was flown into Oakland on May 2, 1946, to become part of the AeroTech Industries school. When AeroTech closed its Oakland campus, it was flown to the school's location at the Glendale Airport. Its final fate is unknown. *William T. Larkins*

After the war, B-24J 42-50801 was delivered to the Boy Scout troop at Urbana, Ohio. In 1951, it was acquired by an aircraft dealer and flown to Dallas Love Field, Texas, where it was refurbished as a cargo carrier. It was registered N299A, and was later sold to Bolivia and registered CP-611. This Liberator was heavily damaged in a 1964 landing accident. It was subsequently rebuilt and given a new registration of CP-787. *Author's Collection*

Air Force's 11th Bomb Group, 98th Bomb Squadron. *Tricky Mickey* returned stateside and was delivered to the Reconstruction Finance Corporation's Storage Depot 41 at Kingman, Arizona, on November 11, 1945. A photo of this aircraft surfaced at an unknown field with a 1947 model Chrysler in the background, so it definitely did not end its days at Kingman. If not Gastonia, where?

After serving the Boy Scouts for a few years, Urbana's 42-50801 and Gastonia's bomber were acquired by an aircraft dealer, made airworthy and flown to Dallas, Texas, for repair and modification before beginning a new career south of the border.

Shortly after the Gastonia bomber was registered to a new owner, the Federal government was not about to let the sale slip by without notice and took the Piedmont Scout Council to court, suing them for $10,000, claiming that the scouts had violated the original sales agreement by selling the aircraft for any purpose other than to reclaim its aluminum. Realizing the scouts could not afford to defend themselves and that the aircraft was long gone, the government dropped its lawsuit in March 1957, and the bomber faded into the citizens of Gastonia's collective memory.

Another mystery that comes into play is whether or not the Gastonia bomber ended up in Mexico registered XA-KUN. Converted into a cargo carrier, this Liberator was operated by Transportes Aereos de Mérida, S.A. (TAMSA), which was partially owned by Pedro Infante Cruz—an actor and singer in Mexico on par with Elvis Presley. On April 15, 1957, Cruz was copiloting XA-KUN when it crashed shortly after takeoff.

War Memorials

In addition to the Boy Scout bombers, only three Liberators served in tribute to the airmen who died during World War II; however, of those three, two were combat veterans.

During the war, the citizens of Joplin, Missouri, bought $300,000 worth of Category "E" War Bonds, which was enough to entitle them to name a bomber. Their choice was Ford-built B-24J 42-50535 that they christened *Joplin Jalopy*. This Liberator began operations on July 29, 1944 with the 506th Bomb Squadron of the 44th Bomb Group flying from RAF Shipdham, England. After completing sixty-six missions, *Joplin Jalopy* returned stateside on May 31, 1945. The bomber was displayed at the Joplin Municipal Airport for a number of years and is believed to have been scrapped by 1950.

Flandreau, South Dakota, received combat veteran B-24J 44-40238 *Thunderbird* for its airport memorial. *Thunderbird* flew with the 852nd Bomb Squadron, 491st Bomb Group at North Pickenham, England, flying its first mission on June 2, 1944. It returned home after the end of the war in Europe, on June 19, 1945. The bomber made its way to the storage yard at Altus, Oklahoma, where it was stricken from the USAAF's inventory on October 26, 1945. The plane came under the Reconstruction Finance Corporation's management and they transferred the bomber to the Flandreau Civil Air Patrol. The Liberator was subsequently transferred to the local Boy Scouts for use as a club house, but the aircraft had deteriorated and suffered at the hands of vandals and was not suitable for their use. *Thunderbird* had been parked near the airport entrance from 1947 to 1957, when it was sold to a scrapper for $1,500.

The Ford plant in Willow Run, Michigan, built more than 8,600 B-24 Liberators, and it was only fitting that the local American Legion Post preserve an example of the locally built bomber. The Edsel B. Ford American Legion Post acquired Ford-built B-24J 44-48834, which had been delivered to the Army Air Forces on August 14, 1944, and flown directly to storage at Reading, Pennsylvania. On April 1, 1945, 44-48834 was flown to Chanute Army Air Field, Illinois, where the aircraft became part of the 3502 Base Unit's training flight. On February 16, 1946, the aircraft was turned over to the Reconstruction Finance Corp. and subsequently flown to the Edsel B. Ford American Legion Post, arriving at Willow Run Airport on February 26, 1946. Three months later, on May 26, a dedication ceremony for the American Legion Post and its newly acquired air battle monument was held, with Henry Ford and his wife as guests of honor. The Edsel B. Ford Post's B-24J sat in a field near the post and, over the years, was heavily vandalized. When the American Legion Post was disbanded in 1950, the aircraft was scrapped.

Aeronautic Trade Schools

Of the four B-24s that went to trade schools, one survived to be rescued and restored by the U.S. Air Force Museum's Heritage program that enabled bases to form their own museums and to collect aircraft representing the base's or the air wing's history.

The sole survivor is B-24J-20-FO 44-48781, which was delivered to the Army Air Forces on August 10, 1944, at Reading, Pennsylvania. This aircraft spent most of the war at Chanute Field, Illinois, and was

In the early 1950s, B-24J 44-48838 was purchased in 1954, and registered N79428. It, too, ended up at Dallas Love Field, Texas, with Southwest Airmotive where it was modified into a freighter. Sold to Bolivia, the plane was registered CP-618 and crashed on May 21, 1960, at El Alta Airport, La Paz Bolivia. *Erwin J. "Pete" Bulban*

PB4Y-1 BuNo 38777 flew with VPB-116 as *Sleepy Time Gal*. After the war it was acquired by Bob Sturges with the intent of putting it on top of a gas station. When Art Lacey beat him to the punch by putting a B-17 on top of his Milwaukee, Oregon, gas station, Sturges abandoned his plans. The bomber was sold to a Bolivian carrier and registered CB-589, later CB-89, and was written off in 1961. *Frederick Johnsen Collection*

Fresh out of the Navy's aircraft storage depot at Litchfield Park, Arizona, B-24M 44-42067/PB4Y-2 BuNo 90232 is seen at the Phoenix Sky Harbor Airport being readied for its flight to South America. The Liberator wears the U.S. registration N5238V, which was crudely applied in white paint. More than a decade later, the Liberator was seen wearing Frigopar titles, and the Brazilian registration PT-AZX, on the ramp at Congonhas Airport, Sao Paulo, Brazil, in February 1970. *Brian Baker/Hélio Higuchi*

transferred to the Reconstruction Finance Corp. at Altus, Oklahoma, on December 7, 1945, and sold surplus to the Spartan School of Aeronautics of Tulsa, Oklahoma, on the same day. The Liberator was used to train mechanics for a number of years and in 1960, the school sold the bomber for scrap. Apparently the scraper took the engines and propellers and abandoned the remainder of the Liberator. It was pushed around the airport for more than a decade until acquired by the National Museum of the U.S. Air Force and moved, by

B-24J 44-48781 was built at Ford's Willow Run plant and was accepted on August 10, 1944. The aircraft had a number of stateside training assignments and was flown to storage at Altus, Oklahoma, on December 7, 1945. It was sold surplus to the Spartan School of Aeronautics in Tulsa, Oklahoma, and was used to teach future airframe and powerplant mechanics their craft. This Liberator is now on display at Barksdale AFB, Louisiana. *Brian Baker*

helicopter, to Barksdale AFB, Louisiana, for restoration and display. The Liberator is now displayed at the air force base's Barksdale Global Power Museum as *Louisiana Belle II*.

Aero Tech Industries had schools in both northern and southern California—Oakland and Glendale respectively—where they trained future aircraft mechanics. A Ford-built B-24M 44-51556 was accepted by the Army Air Forces on April 25, 1945. On June 29, the aircraft left West Palm Beach, Florida, for service in the Panama Canal Zone. The bomber flew patrols from Albrook Field, Panama, until November, when the bomber returned to Morrison Field, Florida, to serve with the 1103 Base Unit. On January 27, 1946, the Liberator was flown to storage at Kingman, Arizona. From Kingman, the bomber was flown to Oakland, arriving on May 2, 1946. The Liberator was used as a training aid by future pilots and mechanics for the next three years. When the school closed in 1949, it was reportedly flown to Glendale to serve at the school's southern California location. From there the trail goes cold.

The Northrop Aeronautical Institute was founded in 1942 by Jack Northrop. In 1946, after the war, the institute began offering classes in a number of aeronautical disciplines. The school acquired a number of aircraft including Ford-built B-24J 44-51842. The bomber was delivered to the Army Air Forces on June 12, 1945, and was shuffled around a number of West Coast and Midwest air bases including Lincoln, Nebraska; Gowen Field, Idaho; and Spokane, Washington. Redesignated a TB-24J, the Liberator was flown to storage at Kingman, Arizona, on December 11, 1945. It subsequently joined the training aircraft at the Northrop Aeronautical Institute, which included a gun-nose B-25H Mitchell (43-5022) medium bomber, P-61C Black Widow (43-8364), and a variety of AT-6s, BT-13s, and UC-78s.

After training students on it for nine years, the school put the aircraft up for disposal. Warbird broker Bob Bean and H. D. Fox won the bid, acquiring the B-25H for $1,500, the P-61C (less engines) for $900, and the B-24J for $2,200. Bean and Fox transported the aircraft to their facility at Blythe, California, where the planes were reduced to parts.

One of the technical trade school Liberators remains unidentified. After the end of the war, a B-24 Liberator was delivered to the Pittsburgh Institute of Aeronautics at the Pittsburgh International Airport in Pennsylvania. It had an unusual nose modification and the interestingly enough, the guns were retained in the upper turret.

The South American Connection

Although the majority of Liberator instructional and memorial airframes were scrapped, a pair of the Boy Scout bombers went on to join a number of other B-24s hauling freight and passengers in Latin and South America. Keeping in mind the contract clause that stated schools and civic groups could not fly their aircraft and that title remained with the U.S. government, how did they start new careers south of the U.S. border?

In the early 1950s, demand for inexpensive cargo hauling aircraft capable of lifting a sizeable payload from hot and high airports like La Paz, Bolivia, were being sought. Aircraft dealer McGee-Ingram, Inc., of Dallas, Texas, had its hand in a number of the Liberators that went south of the U.S. border. Apparently a crew from McGee-Ingram, or more likely Southwest Airmotive, traveled around the country preparing the scout and school bombers for ferry flights to Dallas Love Field, where the planes were modified for their new roles.

The Bolivians acquired five B-24 Liberators, the first of which was an ex-Boy Scout Bomber. B-24J-5-FO 42-50801 had served the Boy

Camera-shy B-24J 42-78770 became XB-RIS when sold to Mexican operator Productos Marinos del Sudeste in February 1952. Note the art deco PMS logo on the forward fuselage. This aircraft's final disposition is unknown. *via Gary Kuhn*

LB-30 AL583 was returned to the United States after serving overseas on the return ferry run, bringing pilots back across the Atlantic Ocean so they could fly more aircraft to England. In RAF service, this aircraft was a Liberator Mk II. It was sold surplus to an American aircraft broker who exported it to Panama in October 1947. It was later sold to a French airline. *via Gary Kuhn*

Scout troop in Urbana, Ohio, and was acquired by 1951. There is still a debate as to whether or not this aircraft was actually delivered to Urbana, or to a larger airfield in the area. Wherever it was located, a crew came in and prepared the Liberator for a ferry flight to Love Field, Dallas, Texas. Here, Southwest Airmotive stripped the bomber and converted it into a freight hauler. The B-24 was registered N299A

B-24M 44-41916 at El Alto Airport, La Paz, Bolivia, in February 1973, wears the registration CP-576. Notice the pile of Liberator remnants in the background under CP-576's tail. The aircraft is undergoing an engine and propeller change, and would soldier on for another two years before being retired. *Hélio Higuchi*

B-24M 44-41916 had one U.S. civilian owner and was sold to a Bolivian airline in 1951. The bomber hauled freight and cattle carcasses over the Andes Mountains until 1975, when it was parked to provide spare parts for other aircraft. Thirty years after going south of the border, the Liberator was recovered by the U.S. Air Force Museum and shipped to the Castle Air Museum in Atwater, California, where it was restored. It is seen here shortly after its arrival at the museum. *Nicholas A. Veronico*

for a short period before being sold to a Bolivian operator. The aircraft was registered CP-611, and began its new career in hauling cattle carcasses in the Andes. The Liberator was involved in a landing accident on March 27, 1964. The plane was heavily damaged and was rebuilt, eventually reregistered as CP-787.

The Bolivians lost two Liberators within about a year of each other. CP-618, B-24J 44-48838, ex-N79428 was written off on May 21, 1960, at El Alta, Bolivia. This was followed by the crash of B-24J 44-0389, which had gone south of the border and became CB-89, later CB-589, and was written off in an accident at Santa Rosa de Yacuma, Boliva on September 17, 1961.

One of the most interesting of the Bolivian Liberators was C-87A 43-30570 (ex RY-1 BuNo 67798) that was sold surplus in early 1946. The plane passed through a number of owners until January 2, 1948, when it was acquired by William "Bill" Odum. The following month, Odum sold the Liberator to Milton Reynolds for his China expedition; he christened the plane *The Explorer*. Odum and Reynolds had broken Howard Hughes's around-the-world speed record in 1947, flying a war-surplus Douglas A-26 Invader. Reynolds believed the mountain K2 was taller than Mount Everest, and planned to use the Liberator to prove that point. Reynolds and Odum did, in fact, fly over K2, only to learn that it was only 28,251 feet tall compared to Everest's height of 29,029 feet. *The Explorer* had a gear collapse and Reynolds called off the remainder of the expedition. Upon returning to the United States he gave the plane to Odum, who in turn sold it to the Johns-Manville Co. for use as an executive transport. The bomber/transport passed to Eivind Rohrt Aviation Supply Co. of Dallas, Texas, and was subsequently sold to Frigorifico Ballnian Ltd., in La Paz, Bolivia (registered CP-75) on March 23, 1951. The plane's U.S. registration was cancelled on May 15, 1951. Flying in the Andes, CP-75 was involved in a landing accident on July 10, 1952. Twelve years later, in January 1964, ownership passed to Compania Boliviano de Aviación, and it was destroyed in a crash on February 8, 1964.

It's interesting to note that Brazil operated a number of single-tail PB4Y-2 Privateers and only a single, twin-tail Liberator. Built as a B-24M at Consolidated's San Diego factory and assigned serial number 44-42067, the aircraft was immediately transferred to the U.S. Navy as modified into a PB4Y-1P photographic reconnaissance aircraft and given BuNo 90232. The Liberator flew with VD-2 and later VPP-2 before being sold surplus. The plane was registered N5238V, and was seen at Phoenix Sky Harbor Airport being readied for the flight south of the border. Frigopar became its operator in Brazil and the aircraft was registered PT-AZX in 1969. It was last reported as derelict in 1975 at Belem, Brazil.

The Chilean air carrier Air Chile acquired a single, ex-Royal Canadian Air Force Liberator (RCAF 574) that had most likely been acquired from surplus Canadian stocks by the Charles Babb Co.

Before this Liberator became known as the Commemorative Air Force's *Diamond Lil*, it was operated by the Mexican state oil company PEMEX as an executive transport. It is seen in the mid-1950s at Minneapolis Airport wearing a Mexican flag and is registered XC-CAY. *Gary Kuhn*

RCAF 574, ex 44-10583, along with five B-24Ms, had been modified into transports at No. 4 Repair Depot, Rockcliffe, beginning on June 30, 1944. RCAF 574 was fitted with a galley and seating for twelve in the main cabin, and began operations on August 30, 1944 flying with No. 168 Heavy Transport Squadron. Here the Liberator transported the Canadian Prime Minister and his staff in RCAF 574. On July 15, 1946, the plane was transferred to No. 9 Transport Group where she operated until being transferred to No. 12 Squadron on April 1, 1948. Two months later, on June 8, 1948, the aircraft was removed from the inventory and put up for sale.

After buying the aircraft, the Babb Co. then sent the Liberator to Dallas for modification by Southwest Airmotive, where it gained

the U.S. registration N1246V. The luxurious Liberator was subsequently sold to Chile where it gained the registration CC-CAN (later CC-CAN-0120). Its final disposition is unknown.

Three Liberators were operated in Mexico. Aside from XA-KUN, XB-RIS went south of the border on February 19, 1952. XB-RIS was a Dallas-built B-24J 42-78770 that was sent to storage at Chanute Field, Illinois, on March 31, 1947. It was sold surplus and registered N1551M. During its career in Mexico, the bomber was operated by Productos Marinos del Sudeste, and the PMS logo was seen on the aircraft's nose. Very few photos of this aircraft exist, and its final disposition is unknown.

Built as a B-24A on order for the British, this aircraft was delivered as an LB-30 with RAF serial AM927. On its delivery flight, the Liberator was involved in a landing accident en route to Canada, from where it would fly across the Atlantic Ocean. Consolidated Aircraft rebuilt it, added the later model's longer nose, and flew it as a company transport. After the war it was sold to the Continental Can Co., then to PEMEX, and in 1967 was acquired by the CAF. *William T. Larkins*

One LB-30 Liberator, RAF serial AL583 was sold surplus at Walnut Ridge, Arkansas, and registered NL4674N to Oceanic Air Transport of New York City. This aircraft was christened *Santa Maria*, and subsequently exported to Panama. The aircraft's U.S. registration was issued on April 15, 1947, and the plane had been exported to Panama before October 29, 1947. It was registered RX-102, and after its service in Panama, was sold to a French airline and reregistered F-OAAD.

Two Came Back

Of the more than twenty B-24 Liberators that went from surplus sale to schools, memorials, and commercial service in South America, only three survive today. From the schools and memorials Liberators, only B-24J 44-48781, the ex-Spartan School of Aeronautics instructional airframe survives at the Barksdale Global Power Museum.

Two Liberators did return from Latin and South America. B-24M 44-41916 was brought back to the United States to become part of the U.S. Air Force's collection. B-24M 44-41916 was delivered to the Navy on October 31, 1944, as BuNo 90165 and served at NAS Hutchinson, Kansas, with an operation training unit (OTU-VB4-1). At the end of the war, BuNo 90165 was transferred to the aircraft pool at NAS Clinton and was stricken on November 30. Sold surplus, the plane was registered N5141N. Five years later, in 1950, it was sold to Salem Engineering Co. of Salem, New York. Its next owner registered it as N4907L. Compania Boliviano de Aviación of La Paz, acquired the bomber on March 22, 1951. In 1966, its ownership was transferred to Bolivian Overseas Airways, also of La Paz. The former bomber hauled cattle carcasses and other goods throughout the mountainous regions of South America until 1975 when it was withdrawn from service to become a source of spares for other Liberators. In 1981, 44-41916 was acquired by the U.S. Air Force Museum. The following year it was shipped to Houston, then on to Castle Air Force Base, Atwater, California. From 1982–1989, parts were gathered and thirty thousand volunteer hours were spent reassembling the bomber. It was rolled out on November 11, 1989, in the colors of the 93rd Bomb Group's *Shady Lady*. Today the bomber is displayed at the Castle Air Museum's air park.

The second Liberator to return from south of the border has probably been seen by more people than any other B-24. The Commemorative Air Force's Liberator was built for the British as an LB-30 serial number AM927 (B-24A 40-2366) and is known as

Diamond Lil. It was the twenty-fifth Liberator built, and on its delivery flight, it suffered a landing accident at Albuquerque, New Mexico, on July 24, 1941. The bomber was rebuilt and flown by the USAAF and later Consolidated Aircraft. Consolidated registered the plane NL24927 and operated it as a company transport until November 1948. Continental Can Co. bought the transport and registered it as N1503. Five months later, in April 1959, AM927 was acquired by the Mexican state-owned oil company Petroleos Mexicanos (PEMEX) and reregistered XC-CAY.

In the mid-1960s, when the founders of the Commemorative Air Force's (CAF) predecessor organization, the Confederate Air Force, were looking to find one example of every World War II type, they had trouble locating an operational B-24 Liberator. In 1968, the CAF was able to purchase the PEMEX transport.

In 2006-2007, the CAF reconfigured the LB-30 to B-24A configuration. This effort was led by then-squadron crew chief Gary Austin. Metal skins covering the tail and tunnel gun positions were removed, uncovering about ninety-five percent of the original structure. In addition, the radio operator's floor was lowered to its original position. The plane received a new coat of paint and was christened *Ol' 927*. After flying the airshow circuit for a year or two, it was decided to put the plane back in its iconic *Diamond Lil* markings. Today this B-24A can be seen on the airshow circuit each spring and summer.

Of the Liberators that went to schools, memorials, and Boy Scout troops, plus those that went south of the border, only three are known to survive, one of which is a flyer. Are there others out there? Are there major components hidden in sheds and barns near where some of the Liberators were scrapped? Only time will tell.

School and Memorial B-24 Liberators (in serial number order)

B-24-1-FO	42-50535	Joplin Municipal Airport	Joplin, Missouri	*Joplin Jalopy*
B-24J-5-FO	42-50801	Boy Scout Troop	Urbana, Ohio	
C-87A-CF	43-30569	Boy Scout Troop	Armonk, New York	
B-24J-150-CO	44-40238	Flandreau Airport	Flandreau, South Dakota	*Thunderbird*
B-24J-20-FO	44-48781	Spartan School of Aeronautics	Tulsa, Oklahoma	
B-24J-20-FO	44-48834	Edsel B. Ford American Legion Post	Willow Run, Michigan	*Old Number 139*
B-24M-30-FO	44-51556	Aero Tech Industries	Oakland, California	
		Aero Tech Industries	Glendale, California	
B-24M-30-FO	44-51842	Northrop Aeronautical Institute	Los Angeles, California	
B-24	unknown	Pittsburgh Institute of Aeronautics	Pittsburgh, Pennsylvania	
B-24	unknown	Boy Scout troop	Gastonia, North Carolina	

Latin and South American Liberators

Latin American Liberators

Registration	Model	Disposition/ Notes	U.S./ RAF serial	
Panama				
RX-102	LB-30		AL583	to RFC/Walnut Ridge on Jan. 15, 1946, ex-NL4674N
Mexico				
XA-KUN	B-24M		unknown	Crashed, Merida, Mexico, April 15, 1957
XB-RIS	B-24J-1-NT		42-78770	to RFC/Chanute Field, March 31, 1947, ex-N1551M, to Mexico Feb. 19, 1952.
XC-CAY	LB-30		AM927	Airworthy with Commemorative Air Force, Midland, Texas

South American Liberators

Registration	Model	Disposition/ Notes	U.S./ RAF serial	
Brazil				
PT-AZX	B-24M-15-CO		44-42067	ex-BuNo 90232, sold surplus, Litchfield Park, ex-N5238V
Bolivia				
CB-76	B-24M-5-CO		44-41916	ex-N4970L, ex-CB-76, currently displayed, Castle Air Museum, Atwater, California.
CB-89	B-24J-160-CF		44-40389	ex-N66569, ex-CB-589, written off September 17, 1961, Santa Rosa de Yacuma, Bolivia
CP-75	C-87A-CF		43-30570	ex-N5151L, ex-CB-575, written off February 8, 1964
CP-787	B-24J-5-FO		44-50801	to RFC/Kingman Jan. 12, 1946. Ex-N299A, ex-CP-611. Crashed March 27, 1964 and rebuilt. Reregistered as CP-787.
CP-618	B-24J-20-FO		44-48838	ex-N79428, written off May 21, 1960, El Alta, Bolivia
Chile				
CC-CAN-0120	C-87		44-10583	ex-RCAF 574, ex-N1264V

Opposite: Is this the Gastonia B-24? Many say no as the Gastonia bomber is reported to have been a C-109 tanker without a nose turret. This aircraft appears to be a memorial, having a sign under its left wing. But where was this photo taken, and where was it a memorial? *Author's Collection*

Above: B-24D *The Squaw* came home and made a war bond tour of aircraft factories and towns around the United States and reportedly ended up in the Tampa, Florida, area as a war memorial. *U.S. Army Air Forces*

The Mystery Liberators

There were other Liberators that went to schools or served as memorials within the United States—some are rumors, and for others there is photographic proof. Can readers fill in the blanks?

B-24	Serial unknown	Pittsburgh Institute of Aeronautics, Pittsburgh, Pennsylvannia
B-24	Serial unknown	Boy Scout troop, Gastonia, North Carolina
B-24	Serial unknown	Nose number "27" (see page 197)
B-24M	44-51901	Location unknown

B-24M   44-50960 *Tricky Micky*

5th Air Force, 90th Bomb Group and 7th Air Force, 11th Bomb Group, 98th Bomb Squadron veteran *Tricky Micky*. To RFC Kingman, Arizona, on November 9, 1945.

B-24D 41-11761 *The Squaw (Sneezy)*

98th Bomb Group, 343rd Bomb Squadron

After the August 1, 1943, raid on Ploesti, Romania, this Liberator returned to the United States for a War Bond tour. Some sources say she became a war memorial at Tampa Bay, Florida, while others report her as scrapped in November 1944.

Combat veteran B-24M 44-50960 served in the Pacific Theater with the Fifth and Seventh Air Forces, carrying the name and nose art *Tricky Mickey*. After the war, 44-50960 was sent to Kingman, Arizona, for storage. It is seen somewhere, a long way from Kingman, with a new Woody under the tail, thus this photo was taken after its stint in storage. This aircraft became a memorial or a technical school airframe, but where? *Museum of Flight Archives*

B-24M 44-51901 sits at a civilian airport after the war. Note that it has radar installed in place of the belly turret. It was probably a war memorial, but where was it and where did it end up? *Museum of Flight Archives*

The Pittsburgh Institute of Aeronautics, a technical school located at the Pittsburgh International Airport, Pennsylvania, received this B-24 for use as a ground instructional airframe. Note that the aircraft was delivered to the school with a pair of .50-caliber machine guns in the upper turret. The aircraft's serial number is unknown, and no photos from an angle that would show the number on the vertical stabilizer have surfaced. *Courtesy Pittsburgh Institute of Aeronautics*

Hidden in Plain Sight
Mustangs on Poles

John Muszala II tucks the NACA Mustang in close over Nevada's Pyramid Lake in September 2013. Known as NACA 127 when in service as an aeronautic research tool, this Mustang served as a high-speed flying test bed. Results from these tests, coupled with what was learned in wind tunnels, and captured German swept-wing research, led to the jet-powered fighters of the late 1940s and 1950s. *Richard VanderMeulen/Photo plane pilot: Brant Seghetti*

North American P-51 Mustangs on Poles? More than seventy years after the type's first flight, a genuine P-51 Mustang on a pole is a crime. Exposed to the elements, they quickly deteriorate from the inside out and can fall victim to vandals. Most pole-mounted aircraft are a curiosity when first seen, but quickly fade into the background, hidden in plain sight. Thankfully, most of the Mustangs on poles in the United States have been rescued, or at least lowered to the ground and are now sitting on their gear, but there are a few skewered Mustangs still sitting out in the weather.

One of the first Mustangs on a pole was P-51A 43-6274 that was parked at the storage depot at Altus Army Air Field, Oklahoma. In 1946, the Allison-powered fighter became a memorial at the Frederick Middle School, Frederick, Oklahoma. It's interesting to note that the school's main building, with the pointed roofs, is still present today, while the domed church in the background has been rebuilt, and the Mustang's former roost is a field with construction now taking place. *Martin Kyburz/Swiss Mustangs collection*

One of the earliest Mustangs on a pole was P-51A 43-6274. This Allison-powered P-51 was accepted by the USAAF on April 26, 1942, and flown to its first assignment at Waycross, Georgia. The following month it was assigned to Memphis, and then to Nashville to serve with the Fourth Fighter Group. In August 1943, P-51A 43-6274 was redesignated a TP-51A trainer and assigned to the 568th Base Unit at Brownsville, Texas. From there the TP-51 flew from Greenwood, Mississippi, until the end of the war. At the conclusion of hostilities in Europe, TP-51A 43-6274 was flown north and stored at Altus Army Air Field, Oklahoma. At Altus, 43-6274 joined more than 2,600 stored B-17 and B-24 bombers along with hundreds of tactical fighters ranging from P-38s and P-40s to P-47s and various other models of P-51s.

Transferred under the Reconstruction Finance Corporation's schools and memorials program in 1946, 43-6274 ended up in front

Wally Erickson had bought the early, high-back Mustang in the 1950s and kept it stored until John Crocker was able to make a deal for the A model. Charles Nichols of the Yanks Air Museum, Chino, California, acquired the Allison-powered Mustang project from Crocker. The project's fuselage arrived on a flat-bed truck in 1978. *Yanks Air Museum*

of a middle school in Frederick, Ohio. Had it not become a war memorial in America's heartland, it would have been scrapped with the other aircraft at Altus.

This P-51A was not on the pole very long, as it was purchased, set on its landing gear, and prepared to compete in the September 1948 National Air Races in Cleveland, Ohio. Although it arrived at the Cleveland race site with Tommy Mason at the controls, 43-6274 did not complete. By 1950, the Mustang had been sold to Harry McCandless and Ben Widfelt, who kept the plane at Council Bluffs, Iowa. Three years later, they sold the fighter to Wally Erickson of Minneapolis, Minnesota, who stored the aircraft for more than twenty years. Erickson later sold the Allison-powered Mustang project to John Crocker, who, in 1978, sold it to Charles Nichols of the Yanks Air Museum at Chino, California.

Once at the Yanks Air Museum, 43-6274 was put into the restoration queue and its rebirth was guided by the late Stan Hoefler. It took more than a decade to finish, but when it emerged, Hoefler's attention to detail was obvious, as 43-6274 looked pristine. The Mustang was rebuilt as an F-6A photo-reconnaissance Mustang in the colors of the 67th Recon Group. All of the photographic equipment has been installed in the aircraft, including the large format, side-facing camera behind the pilot's seat. This restored A-model Mustang is displayed at the Yanks Air Museum's Chino, California, facility.

D-Model Mustang on a Pole

Another Mustang that sat on a pole, for far longer than 43-6274, was Dallas, Texas-built P-51D 44-84900, which was delivered on September 4, 1945. This aircraft flew from the factory to Brookley Army Air Field, Alabama, before it was requisitioned with some other D models by the National Advisory Committee for Aeronautics (NACA, the predecessor of today's National Aeronautics and Space Administration, NASA) and flown to their facility at Langley, Virginia. P-51D 44-84900 lost its military identity and became "NACA 127" while flying as an airborne testing laboratory.

After a decade-long restoration, 43-6274 emerged as an F-6A photo reconnaissance Mustang wearing the colors of the 67th Recon Group. Notice the camera looking through the rear window of the cockpit. *Nicholas A. Veronico*

NACA's most obvious external modification to the Mustang is its tall vertical tail that increases lateral stability during high-speed dives. The aeronautics agency also fitted the top of the outer wing panels with model test sections. The test sections are painted white and located outboard of the wing gun bays.

"Pilots would notice during high-moisture days that they could see the flow along the upper surface of the wings, during high speed pull-ups," said William C. "Bill" Allmon, owner of NACA 127. "They determined that if they took the plane up to a certain altitude, did a high-speed dive and a 4g pull-up, they could get the flow over the wing into transonic speeds. It was basically a flying wind tunnel."

NACA Langley researcher Robert R. Gilruth, then-chief of the Flight Research Section, developed "wing-flow method" testing in 1944, and a number of NACA-operated Mustangs were outfitted for this special mission. Some had cameras that recorded the airflow over and around the airfoils while NACA 127 had cameras mounted inside the fuselage and wings to photograph a variety of instruments reading data from strain gauges. The strain gauges were measuring pressures as transonic airflow passed over the models mounted on top of the wing. Some models were airfoil shapes while others were semi-span aircraft (essentially half an aircraft split lengthwise and mounted flush with the testing surface).

Factory-fresh P-51D 44-84900 is shown as delivered to the National Advisory Committee for Aeronautics' Langley Memorial Aeronautics Laboratory in 1945. The aircraft has not yet been modified for NACA service. The Mustang wore NACA serial number 127 during its testing days. *NASA*

A NACA technician checks the strain gauges inside the wing that will monitor airflow over the semi-span model mounted on the Mustang's wing test section. That model bears a strong resemblance to the Bell X-1. Notice the vane on the test section in the lower left of the photo. *NASA*

On each wing, outboard of the test model is a small square airfoil that protrudes into the airstream. Both of these airfoils provide yaw indications to an instrument, similar to a pilot's directional indicator (PDI), to tell the pilot when the aircraft is being flown perfectly straight. Only when there was no yawing to either side could the test begin and the pilot and engineers would know that the test was performed correctly and that the data collected was valid.

Typically a test pilot would take the aircraft and its experiment up to 35,000 feet and then push over into a dive reaching Mach .71 to .73. Descending to 10,000 feet, the pilot would begin a 4g pull-out while recording data. Then the pilot would climb up to 35,000 feet and perform the test again, and again, and again. At Mach .71 to .73 in a 4g pull-out, the airflow over the wing model would, locally, go supersonic. Capturing how the airflow reacted as it passed around the airfoil model and how it accelerated from subsonic to supersonic flow was, at the time, extremely difficult to model in a wind tunnel. Wind tunnel testing as well as "wing-flow," "drop-body" (airfoil and aircraft shapes mounted on the nose of a bomb dropped from a B-29 at 30,000 feet), and "rocket model" (air foil shapes mounted to the nose of a rocket) tests combined with captured German aeronautical information provided the data used to design America's post-war generation of jet aircraft.

In 1952, 44-84900 was released from NACA service and acquired by the Pennsylvania Air National Guard. The guard used the aircraft as a fighter for a number of years having reinstalled much of the armament. The Mustang was then grounded for use as an instructional trainer. Many Pennsylvania ANG mechanics in the mid-1950s honed their skills on this tall-tail Mustang. In 1973, it was mounted on a pole and displayed at the Greater Pittsburgh Air National Guard Base as a gate guardian. Over the years 44-84900 was displayed in an overall bare metal finish, and was later painted silver with ANG markings. In the late 1980s, while still on the pole, the Mustang was given a World War II combat paint scheme with fuselage codes AJ*S.

While the Mustang was on the pole, the rumor got started, and stuck, that 44-84900 was "the only P-51 to ever land on an aircraft carrier." A Mustang did, in fact, land on an aircraft carrier; however, P-51D 44-8490 was built too late to participate in those trial flights. The testing was known as "Project Seahorse," and Lt. Robert M. "Bob" Elder flew P-51D 44-14017, redesignated an ETF-51D, for this program. During the war, Elder had been awarded two Navy Crosses, one for his actions at the Battle of Midway for dropping a bomb squarely onto the deck of a Japanese carrier in the face of withering anti-aircraft fire while flying a Douglas SBD Dauntless divebomber.

For the Mustang arrested landings, Elder flew more than 150 simulated carrier landings at Mustin Field, near the Naval Aircraft Factory at Philadelphia, Pennsylvania, before taking the fighter aboard ship. On November 15, 1944, Elder demonstrated the Mustang's ability to operate from an aircraft carrier by making four takeoffs and four landings aboard USS *Shangri-La* (CV-38). It was, however, not a task for an inexperienced pilot as the Mustang's stall speed was 82 mph and the arresting hook could not be engaged at speeds of more than 90 mph. Although NACA 127 was also designated an ETF-51D, these facts should lay to rest the rumor that Mustang 44-84900 ever went aboard ship.

Back in the Air, But Not on a Pole

Bill Allmon started flying gliders as a young man, and he raced them for a long time. Using a private plane in his business, he transitioned into more complex aircraft as his ratings evolved. Allmon purchased a Beech Starship as his business grew, but for personal pleasure he had

P-51D 44-84900 with the NACA tall-tail modification to reduce yaw. The aircraft is in service with the Pennsylvania Air National Guard. Note that all of the external NACA test gear has been removed and the fuselage bears the remnants of a buzz number under the cockpit. *via A. Kevin Grantham*

always wanted a P-51 Mustang. "I had decided I wanted a Mustang and looking around the market, any of those nice ones were starting to top $1 million at the time," said Allmon. "I couldn't afford that. I just couldn't. I could, however, afford to cash flow a restoration. That's kind of the way I started. I contacted Mark Clark as he represented a couple of airplanes and they were too expensive, but this project intrigued me because he said it actually had a history."

Allmon went back east in September 1993 and saw the project Mustang at the Air Heritage Inc. museum in Beaver Falls, Pennsylvania. "It was pretty rough looking when I first saw it," Allmon said. The Mustang project sat in the corner of the museum with the wings sitting side-by-side, the tail was off, and the canopy was slid all the way back.

"David Tallichet had made a deal with the Air Force to acquire the plane," Allmon recalled. "I think he traded them a Fiberglas replica in exchange for the aircraft." Tallichet never registered the airplane, and had not done any work to it when his restaurant empire suffered some financial setbacks and he was forced to sell off part of his 120-plus

A little later in its ANG career, 44-84900's markings have been updated. Notice the "0" preceding the serial number denoting an aircraft that is more than ten years old. *Bill Allmon Collection*

aircraft warbird collection. Allmon was interested in the project and a deal was made. After arranging payment, Allmon received clear title to the Mustang from the Air Force. In March 1994, the P-51D project was trucked to John Muszala's Pacific Fighters at Chino Airport, California. Restoration work got underway, and then the project was moved to Idaho Falls in 1996 when Muszala relocated his shop.

"The thing that did the most damage to the airplane was just sitting. It sat out in the weather for a long time. Probably thirty-plus years," Allmon said. "Certainly from the lower longeron down it was a badly corroded mess. I think that's what motivated the Air National Guard to get rid of it.

"All the rot was happening down below and it was starting to affect the structural integrity of the aircraft where it was attached to the pole. The higher up you went in the airplane, the better things were, but all that low stuff, the lower longerons were like a croissant. They were forged in layers and you could actually peel the metal away. The scoop of the doghouse had to be totally redone. The belly skin was gone. The intake trunk was gone; it was corroded, totally eaten away. When we opened it up it, we found that every creature that could get into that airplane had lived in it. It was a nest for everything."

The Mustang sat on the pole with its original engine, but at some point the propeller hub had been changed and the blades were swapped out. Mustang propellers are expensive and one had to be tracked down.

P-51D 44-84900, albeit missing the last digit in this paint scheme, was retired to a pole to serve as a gate guardian at the Greater Pittsburgh Air National Guard Base. Years of sitting outside in the harsh Pennsylvania winters took their toll on the Mustang and it was removed from the pole when the aircraft's structural integrity became an issue. *via Earl Holmquist*

The restoration started out as a rebuild to combat configuration, but as Muszala's team got deeper and deeper into the aircraft, they kept finding the unique fittings of the NACA modifications. This aircraft had never been civilianized and the decision was made to switch course 180 degrees. All of the recently installed armament was removed and the Mustang was brought back to NACA configuration. This was one of the standard-setting Mustang restorations as Muszala's team photographed nearly every part that was taken off, both externally and internally. "In doing that, we found the inside of the skin was marked with doodles and messages and check marks or sign-offs, not just inspection stamps, done by employees at the factory," Allmon said. "Muszala's team took great pains to put all that back the way it was when we found it. We photographed each one of them and carefully recreated them. Now, they do it a lot. Midwest Aero does it and a few other people do it, but this was the first airplane that went through such a detailed restoration process."

The restored cockpit of 44-84900 includes all of the testing equipment when flown in NACA service. *Pacific Fighters*

When the Mustang was originally built, all of the parts were painted before they were assembled, so it was totally bare-riveted throughout. "When we were taking it apart, we were meticulous at putting those colors back on each rib—even though it's a pain in the ass to do—just so it's period perfect," Allmon said. "It really looks like it came off a factory line." One good example is the engine mount. When the cowls are removed, the engine mount is in four different colors, but that's how it came from the factory. The airplane was restored exactly the way it came apart and that's how it went back together.

"During the restoration, we had tried real hard to get information out of NASA and initially they were pretty closed-mouth about things," Allmon said. "They didn't supply much, but, as the restoration went along and, I guess, at some point they started to feel like I was serious about preserving it as a test airplane, they became a lot more forthcoming. I was working with a gentleman out of Virginia, and he got in touch with one of the engineers. One thing led to another and the engineer volunteered a wing flow model, one of the actual test articles. His experiment is the airfoil you see on the left wing. He gave me his test pictures as well.

"I later met NACA test pilot Bob Champine who flew this aircraft. In talking to Bob, they didn't fly the airplane much. He said that the once a mission was flown and the experiments they needed were verified as having been done correctly, they just parked it until the next mission. When another experiment was planned, they would then go out and prepare the aircraft, so they didn't accumulate many flight hours."

Continuing the NACA/NASA connection, during the restoration 44-84900's dataplate was flown aboard the Space Shuttle *Discovery* on

STS-85, a mission that lasted from August 7–19, 1997. STS-85 was the Space Shuttle program's eighty-sixth flight, and *Discovery* cruised above the Earth at an altitude of 168 miles and reached a maximum speed of 17,610 mph on that mission. That makes 44-84900's dataplate the fastest Mustang dataplate—ever!

Muszala's team worked on the Mustang for three years and had it ready to fly in July 1998, just a few weeks before EAA AirVenture at Oshkosh, Wisconsin. Here it won the Post-World War II Grand Champion award and Muszala's Pacific Fighters took home the coveted Golden Wrench Award.

In September 2010, 44-84900 was entered in the National Aviation Heritage Invitational Trophy competition, which is held in conjunction with the National Championship Air Races at Reno, Nevada. Here the NACA Mustang attracted a great deal of attention and garnered top honors, receiving the Rolls-Royce Aviation Heritage trophy. This perpetual trophy resides at the Smithsonian Institute's National Air and Space Museum—Udvar-Hazy Center in Chantilly, Virginia, and Allmon's name is engraved as the 2010 winner. In September 2013, Allmon, Muszala, and the NACA Mustang were invited back to the National Aviation Heritage Invitational as a returning champion where the unique Mustang was a crowd favorite.

"I've put about 350 to 360 hours on it," Allmon said. "And recently I've taken some of the NACA equipment out of the plane so I can put a bag or two in it, and we've put the back seat in it. We can always change it back to the full NACA configuration pretty easily. It takes a couple days to do, but we made it easy to pull all that standard stuff out and put all the NACA stuff back in."

The cockpit is fairly stock with modern radios hidden behind one of the NACA panels. A set of Collins Pro-line radios and a transponder are grouped onto a plate that can be swapped with the original armament panel low in the cockpit behind the control stick. Using butterfly nuts, the plate with the radio heads and the original armament panel can be changed out in a matter of minutes.

Although 44-84900 sat on a pole for more than thirty years, sitting there kept the aircraft from being civilianized. The internal modifications found during the restoration enabled the aircraft to be brought back to the days when NACA used diving aircraft to replicate airflow over wing models at transonic speeds. P-51D-25-NT 44-84900 is a true time capsule of aviation history. When you consider that the Yanks Museum's P-51A and NACA 127 survived because they sat on poles, planes on poles are not all bad and make for lucky finds.

Above: The wing flow model on the left wing of NACA 127's test section is a genuine test shape from the late 1940s. This attention to detail makes this Mustang a crowd favorite. *Nicholas A. Veronico*

Right: NACA 127's Bill Allmon, right, receives the Rolls-Royce Aviation Heritage Trophy at the National Aviation Heritage Invitational competition at Reno, Nevada, in September 2010. The competition is held in conjunction with the National Championship Air Races. The trophy has since been renamed in honor of astronaut Neil A. Armstrong, and is displayed at the Smithsonian Institute's National Air and Space Museum's Udvar-Hazy Center in Chantilly, Virginia. *Nicholas A. Veronico*

Top: The NACA Mustang's D-model lines and the tall tail plus the wing test sections and extended pitot tube are best seen in the air. This angle shows the test shape and each wing's yaw vanes to advantage. *Richard VanderMeulen/Photo plane pilot: Brant Seghetti*

P-51D 44-73972 is one of the lucky Mustangs having been brought down from its pole. The rare fighter is now on its gear at the California Air National Guard's base at the Fresno Air Terminal in central California. *William T. Larkins/Nicholas A. Veronico*

Mustangs on Poles

P-51D 44-72948	West Virginia Air National Guard, Charleston	NLS
P-51D 44-72989	Volk Field, Wisconsin	SS
P-51D 44-73972	California Air National Guard, Fresno	NLS
P-51D 44-74407	North Dakota Air National Guard, Hector Field	NLS
P-51D 44-72123	San Isidro, Dominican Republic (ex-55th FG)	SS
P-51D F-303*	Jakarta-Halim Perdanakusuma International Airport	SS
P-51D F-347*	Armed Forces Museum, Jakarta, Indonesia	SS
P-51D F-363*	Malang AFB, East Java	SS
P-51D FAG-336*	Guatemalan Air Force, La Aurora Air Base	SS

NLS = No Longer Skewered, lowered from pylon/poles and now on its gear

SS = Still Skewered, outside on a pole

* = USAAF serial number unknown

World War II Privateer
Survived as a Firebomber

Flying up the coast approaching the Golden Gate are four of the U.S. Coast Guard's seven P4Y-2G Privateer patrol bombers, from left BuNo 66300, 66302, 66304, and 66306. Note the painted wing tips. These aircraft were virtually new when acquired from surplus Navy stocks held at the aircraft storage facility at Litchfield Park, Arizona. Most of the Coast Guard Privateers had less than ten hours total time when transferred to the sea service. *U.S. Coast Guard via Todd Hackbarth*

When it comes to warbirds, the Consolidated Vultee PB4Y-2 Privateer is a real sleeper. The type has extensive combat history, starting with the war in the Pacific, but its service was eclipsed by the Navy version of the B-24 (PB4Y-1) and the B-29 Superfortress. Privateer squadrons operated from Tinian, the Philippines, Palau, and later Okinawa, to attack Japanese land and sea targets. The Privateer's ability to cruise

World War II-era shot of a fully armed Consolidated PB4Y-2 Privateer in flight. All gun positions are power-operated turrets—bow, two upper, two waist, and the tail. There was no need for a ball turret in this patrol bomber because the bow and waist turrets could rotate down past ninety degrees and their fire would converge below the aircraft. There were instances where a PB4Y-2 was flying very low strafing a Japanese ship and the bow turret gunner was firing with the turret fully rotated down and his rounds ricocheted off the water, striking the rear of the aircraft. *U.S. Navy*

for long distances, use its search and electronic countermeasures equipment, coupled with its heavy armament, enabled the aircraft to be effective at sending ships to the bottom and to dogfight Japanese aircraft when necessary. U.S. Navy PB4Y-2s were responsible for sinking more than 570 ships, damaging another 620, while downing more than forty enemy aircraft at a cost of only fifty-six Privateers.

The U.S. Navy entered the Cold War on April 8, 1950, when four Soviet Lavochkin La-11s flying from Kaliningrad, Soviet Lithuania, intercepted Privateer BuNo 59645 from VP-26 at 12,000 feet near the port of Liepaja, Latvia. After being fired upon by the LA-11s, the aircraft crashed and burned, or exploded in the air, approximately thirty-five miles west-southwest of Latvia. The remains of the crew were never located.

Later in 1950, Privateers were called up for action in the skies over Korea, often dropping flares for Navy fighters attacking truck convoys carrying supplies to North Korean troops. In the early and mid-1950s, the Nationalist Chinese used the type for bombing, sea patrol, and to drop agents into mainland China. In the latter half

BuNo 66302 is seen at Midway Island, early in its Coast Guard career. Note that the aircraft retains all of its Navy electronic warfare antenna domes on the lower forward fuselage. *U.S. Coast Guard via Todd Hackbarth*

Left side view of BuNo 66302 at Coast Guard Air Station San Francisco after July 1957. When the patrol bomber was overhauled in July 1957, all of the antenna domes were removed and the markings reapplied. Notice the different position of the Coast Guard emblem; the star and bar has been moved forward, and a "Rescue" arrow has been added pointing to the pilot's side window. *William T. Larkins*

of the decade and into 1960 and 1961, the French Aéronavale used Privateers in North Africa and Vietnam. French Privateers patrolled the Gulf of Tonkin and flew alongside surplus Grumman F6F Hellcats and F8F Bearcats, as well as Vought F4U-7 Corsairs, supporting the beleaguered garrison at Dien Bien Phu during that April 1954 action.

The Privateer most likely holds the distinction for being the last World War II four-engine bomber-type to have been lost in aerial combat. Nationalist Chinese Privateers were regularly dropping supplies to Kuomintang guerillas in Burma, and on February 14, 1961, the Burmese forcefully put an end to the over-flights. The Burmese sent three Hawker Sea Furies from 1 Squadron, 501 Wing to intercept what they thought was a C-46. The aircraft turned out to be a Nationalist Chinese Air Force PB4Y-2, which was shot down by Pilot Officer Noel Peters, flying Sea Fury FB.11, serial number UB466. Peters's aircraft was struck by return fire from the patrol bomber and he perished when his Sea Fury crashed and burned.

Although the Privateer was overshadowed by new and faster jet aircraft and newer generations of patrol bombers, there is no other four-engine patrol bomber that has such an extensive combat record, nor has one fought with so many different air arms.

Military and Civilian Life

In spite of the Privateer's long and varied combat career, in the warbird market the type has been overlooked by most collectors. Of course their size and four-engine complexity makes them expensive to operate and far more involved than a single-engine fighter or a light attack bomber, but they are true warbirds. In 2011, a PB4Y-2 returned to the airshow circuit, becoming a real curiosity to show veterans and the general public alike. That aircraft is the former Coast Guard P4Y-2G Privateer, BuNo 66302.

During World War II, the U.S. Navy was able to obtain a number of B-24 Liberators to fill its need for a long-range anti-shipping patrol aircraft. On April 6–8, 1943, representatives from the U.S. Navy's Bureau of Aeronautics met with engineers from Consolidated Vultee (later Convair), to discuss a quick modification that would make the four-engine land bomber more suitable for the naval mission. The modifications would include radar, a nose turret, electronic warfare capabilities, and a different crew-seating configuration.

On May 8, 1943, the Navy issued a letter contract for three prototypes and 660 production aircraft ($108.9 million or $165,000 each),

plus $16.3 million worth of spare parts. By war's end, the Navy had ordered 1,370 Privateers. In a patrol/reconnaissance configuration, the aircraft was designed to have a 62,000 pound gross weight, with a maximum speed of 237 mph at 13,750 feet, a fully loaded stall speed of 101 mph, a service ceiling of 20,700 feet, and be able to take off and clear a fifty-foot obstacle with a 4,700-foot take off roll. The patrol bomber had a range of 2,800 miles.

This new naval reconnaissance aircraft, the PB4Y-2 (PB—Patrol Bomber, model Four, Y—Consolidated Aircraft Corp., design two) was fitted with four 1,350-horsepower, supercharged R-1830-94 engines for improved, low-level performance, and six turrets, each fitted with two .50-caliber machine guns each (nose turret, two top turrets, two side blister turrets, and a tail turret). The standard B-24 fuselage was lengthened by seven feet forward of the wing to accommodate the pilots, a radar operator, radio operator, navigator, and an additional top turret. Under the fuselage, behind the nose wheel, an APS-15 search radar in a retractable dome was installed and deployed in flight. The B-24 Liberator's twin tails were removed

Coast Guard P4Y-2G in action showing a Privateer flying overhead after escorting this Navy P2V Neptune to a safe landing at the Half Moon Bay Airport, on the coast near San Francisco. The Neptune was returning stateside when its starboard R-3350 engine had to be shut-down—notice the feathered propeller. Coast Guard Privateers were often called upon to shepherd aircraft in distress flying between the Orient and the U.S. mainland. *Reg McGovern*

and the Privateer was fitted with a single, tall, vertical tail and rudder.

The first squadrons to receive the PB4Y-2 were VPB-118 and VBP-119, which began transitioning to the Privateer in August 1944. VPB-118, the "Old Crows," became the first squadron to take the type into combat, deploying to Tinian Island, in the Marshalls Group, on January 6, 1945. In March 1945, VPB-109, -123 and -124 were selected to take the radio-controlled "Bat" bomb into combat. On 23 April 1945, VPB-109 made the U.S. Navy's first Bat missile strikes of the war against shipping in Balikpapan Harbor, Borneo.

After the Japanese surrendered on September 2, 1945, nearly all of the Navy's aircraft contracts were canceled; however, production of the PB4Y-2 continued, albeit at a slower rate. Approximately half of the order was canceled, but a total of 739 PB4Y-2s were delivered plus an additional 34 RY-3 cargo-carrying versions. The Privateer was built at Consolidated's San Diego, California-factory, and those built after the end of the war were flown to the modification center at Litchfield Park, Arizona, completed for service, and then towed across the field and placed into storage.

PB4Y-2 Privateer BuNo 66302 was accepted on September 10, 1945 and delivered on November 5, 1945. When it was placed into storage a Litchfield Park, the patrol bomber had accumulated four hours of flying time. Seven years later, on June 19, 1952, BuNo 66302

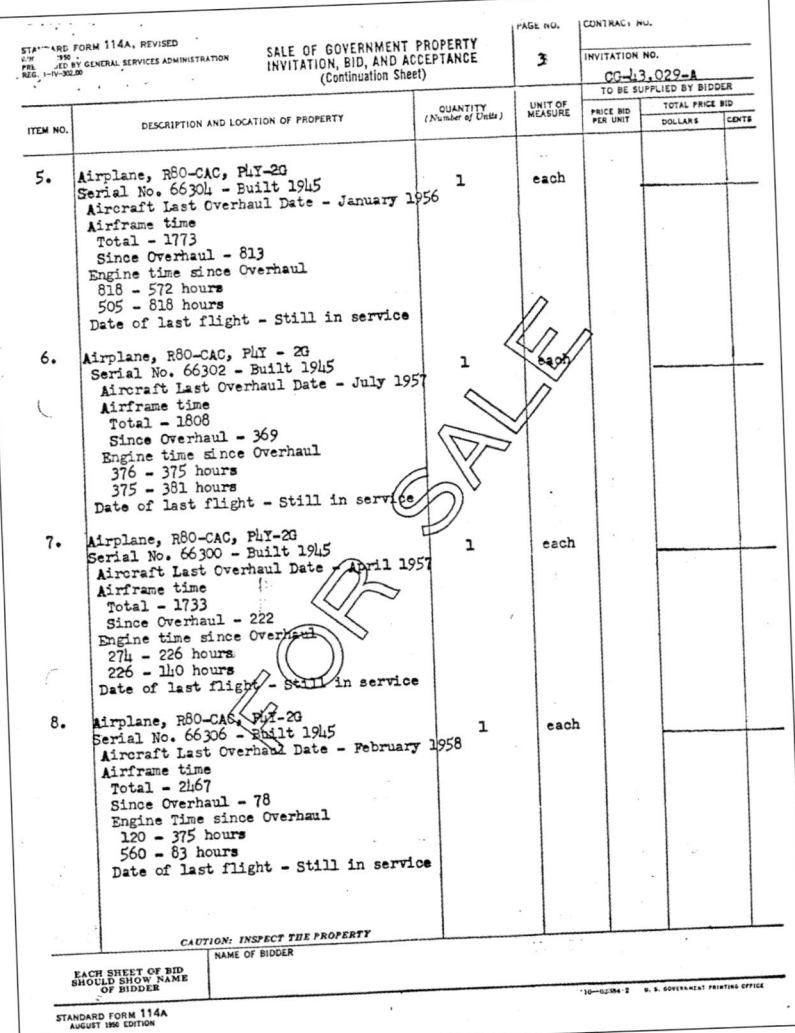

The U.S. Coast Guard offered its Privateer and Flying Fortress fleet for sale while the aircraft were still in service. BuNo 66302 had flown 1,808 hours at the time of the sale, almost all of that time with the Coast Guard. *Scott Thompson collection*

was flown west where it was overhauled at Naval Air Station North Island, outside San Diego. After overhaul, it was stricken from the Navy's inventory with eight hours total time.

In 1951, the Navy overhauled its aircraft designation system and the PB4Y-2 became the P4Y-2. Shortly after the designation change, the U.S. Coast Guard acquired seven Privateers for its long-range search mission. Those aircraft were BuNos 59688, 59876, 66260, 66300, 66302, 66304, and 66306. The Coast Guard Privateers were redesignated as the P4Y-2G and flown to NAS North Island for modification. Here the turrets were removed and replaced with observers' stations—a chair behind a giant window, which had to be the best seat in the service.

The Privateers could not drop life rafts to sailors in distress, but they could arrive in a search area very quickly and help direct aid to stricken vessels. They were also used extensively to escort aircraft that might be having mechanical or navigational problems when making a trans-Pacific flight. For example, during the mid-1950s, the Boeing Stratocruiser was having propeller blade problems, and a couple of aircraft had to ditch between Hawaii and the West Coast. Coast Guard Privateers were stationed at San Diego and San Francisco, California, and Barbers Point, Hawaii, with temporary duty stations at Wake and Midway Islands.

In June 1958, while the aircraft were still flying patrols, the U.S. Coast Guard put its fleet of PB-1s (B-17s) and P4Y-2Gs up for auction. BuNo 66302 was last overhauled in July 1957 and had only 1,808 hours total time on its airframe, 369 since overhaul. At the time, the powerplants were very low time, with 375, 375, 376, and 381 hours respectively on numbers one through four .

Ace Smelting was an aircraft recycling business that had a yard adjacent to the Navy's storage facility at Litchfield Park. Ace Smelting made a lot of money buying surplus Privateers, removing the patrol bomber's four R-1830-94 radial engines, and reselling those to companies upgrading DC-3s and C-47s. Ace Smelting and its competitor Allied Aircraft, could supply two DC-3s per Privateer, and many of the former patrol bombers could be seen sitting tail down/nose in the air, minus engines, waiting to be scrapped.

On November 6, 1958, Ace Smelting won the bid for 66300, 66302, and 66404, having acquired 66260 on February 14, 1958. The individual aircraft sales prices are not known as they were blacked out on the FAA's copy of the bills of sale. Two of the other Coast Guard Privateers were also sold surplus: 59876 was sold to Lysdale Flying Service of South St. Paul, Minnesota for $6,600 and 66306 went to Christler and Avery of Greybull, Wyoming for $10,000. The seventh Coast Guard P4Y-2G, BuNo 59688 was damaged beyond economical repair on June 19, 1957, when its starboard landing gear collapsed while rolling out after landing at CGAS Barbers Point.

Ace Smelting turned its Privateers around for a quick profit, selling 66260 to Big Piney Aviation of Roy, Utah, and, on November 20, 1958, the remaining three to Christler and Avery. On January 29, 1959, BuNo 66302 was registered N2871G, and modified into an aerial firefighting aircraft with the addition of retardant tanks, new cockpit windshields, and up-rated R-2600 engines. The aircraft fought fires for a decade before being sold to Hawkins and Powers Aviation, also of Greybull, on July 7, 1969. Through the years, the plane wore tanker number B21 ("B" for the U.S. Forestry Department region) and later wore tanker number 121. The Privateer air tankers fought fires from Alaska to California and throughout the Western United States for more than forty years.

With Hawkins and Powers, Tanker 121 soldiered on until July 18, 2002, when a sister ship, Tanker 123 (BuNo 66260, N7620C) flown by Ricky Schwartz and Milt Stollak, crashed while fighting the Big Elk fire outside of Estes Park, Colorado. According to the National Transportation Safety Board's April 23, 2004, Safety Recommendation, the aircraft:

> ... experienced an in-flight separation of the left wing while maneuvering to deliver fire retardant over a forest fire near Estes Park, Colorado.
>
> Metallurgical examination of the left wing revealed that it failed at the wing-to-fuselage attachment point along the forward lower spar cap. A fatigue crack measuring approximately twenty-one inches had propagated from the lower portion of the forward spar cap members upward into the spar web. The crack initiated in rivet holes used to attach three "L"-shaped spar cap members to the spar web.
>
> The portion of the wing containing the fatigue crack was obscured by the retardant tanks and would not have been detectable by an exterior visual inspection. The Safety Board determined that the probable cause of this accident was the in-flight failure of the left wing due to fatigue cracking in the left wing's forward spar and wing skin.

The crash of N7620C, and the wing separation and resulting crash of firefighting C-130A N135FF, caused all large ex-military air tankers to be grounded. The NTSB reported that the large, ex-military type air tankers typically underwent flight loads of -0.5 to +3.9gs when pulling out after dropping a retardant load. Any piece of metal exposed to these types of strains for more than forty years will be affected by the stress and suffer some type of fatigue. It's only a matter of when.

When the ex-Coast Guard Privateers were sold surplus, the planes' new owners did not have to strip out too much military equipment as most of that work had been done for them. The Coast Guard removed all of the gun turrets and electronic countermeasures equipment to provide large windows for observers. The air tanker operators did change the cockpit to a one-piece windshield and swapped the Privateer's R-1830-94 engines for more powerful, more plentiful, and cheaper to maintain and acquire R-2600 powerplants. *Milo Peltzer*

The subsequent loss of work for ex-military air tankers impacted Hawkins and Powers and forced the company to liquidate its assets at auction on August 23 and 24, 2007. The sale put two Privateers, a C-82, a number of C-119s and P2V Neptunes, and an A-26 into the warbird market. At the time of the auction, four Privateers sat on the tarmac: BuNo 59701 (N6884C, the second Privateer to wear that registration, Tanker 127), BuNo 59882 (N7962C, Tanker 126), BuNo 66300 (NN2872G, Tanker 124), and BuNo 66302 (N2871G, Tanker 121). BuNos 59701 and 59882 were held out of the auction and retained by Hawkins and Powers. BuNo 66300 was sold to the Yanks Air Museum in Chino, California. After more than six hundred man hours, the Yanks Privateer was made ferryable and delivered to the museum by air on November 22, 2008.

New Life for BuNo 66302

BuNo 66302 was acquired by a group of warbird enthusiasts operating as 4Y-2 LLC, based in Phoenix, Arizona. When acquired from Hawkins and Powers, BuNo 66302 had 7,192.83 hours on the airframe.

Before the aircraft was moved, 4Y-2 LLC had to inspect and repair as necessary any spar damage they found in 66302's wings. "The actual fix is pretty straightforward," said Joe Shoen of 4Y-2 LLC. "What they do is remove a whole bunch of skin where the wing mates with the fuselage, remove all of the doublers, and then peel back the skin approximately two feet. It is unriveted and methodically rolled back. Then the interior of the wing is exposed showing the actual piece with the structural issues and its only one piece that has a problem. They basically put doublers, between eighteen- and twenty-four-inches long and about two-inches wide with L-brackets on either side, to strengthen the spar. They use a specific rivet pattern to reassemble it.

"I'm confident that every plane they ever flew as a fire bomber had some version of this problem. Ours was clearly separating. There was no question, and once it was all taken apart it was easy to see."

As part of the engineering of the spar fix, 4Y-2 LLC worked with the FAA to develop an inspection and monitoring program for the wing spar area. This program will culminate in a five-year inspection involving a nearly complete tear-down of the area. "The crew that did our spar repair were many of the same people who worked at Hawkins and Powers," said Shoen. "What I observed is that they were all struck by the tragedy of the crash of Tanker 123, and so committed to not having a repetition of that accident. They gave the kind of quality one would want if you were going to have your kids flying on the plane. They just knocked this aircraft apart, taking all of the control surfaces off the plane, every inspection port, then they inspected, repaired, and corrosion-proofed each part. It was absolutely the right thing to do, and it has extended the life of the plane considerably."

At the Controls

Chief pilot on the P4Y-2G is Woody Grantham. He was the "G" in T&G Aviation, an aerial application company that once owned PB4Y-2 BuNo 59819 that is now under restoration at the Lone Star Flight Museum in Galveston, Texas. Grantham is one of five or six pilots still flying that have between three to four thousand hours in type.

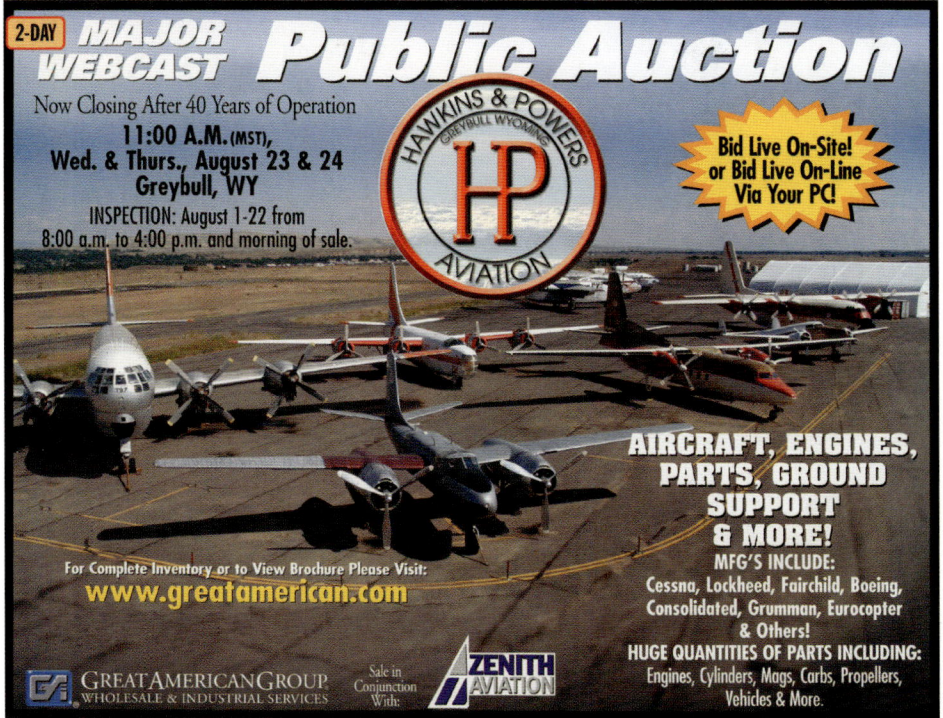

After the crash of a fire fighting C-130 and a Privateer, the U.S. Forestry Service grounded all large, ex-military air tankers. The grounding forced a number of air tanker operators out of business, and Hawkins and Powers suffered this fate. The company's aircraft were put up for auction, and Bunos 66300, Tanker 124, and BuNo 66302, Tanker 121, were the stars of the sale. *Milo Peltzer collection*

"It's a nice airplane to fly. It's an old style airplane in that none of the controls are boosted," said Grantham. "You do everything manually, the rudder, elevator, and the ailerons are all controlled by cables. The controls are fairly heavy."

The ex-fire bomber's R-2600 engines give it more power than the original R-1830-94s. With the R-2600s, the patrol bomber is about ten knots faster and is able to maintain more power at high density altitudes—a good thing when operating out of Arizona's hot and high airports.

"I was involved with one of the companies that did the supplemental type certificate to convert the Privateer over to Wright R-2600 engines," said Grantham. "We didn't do any engine to wing re-engineering. We limited the R-2600's operational maximum horsepower output to the maximum of the old R-1830 engines. In other words, at sea level if we're taking off, we use about two-thirds the throttle. What's nice about it is if we're at a high density altitude airport, say at the Grand Canyon in Arizona or places up in the northwest, we can just push the throttles further forward and get the same manifold pressure out of it. It gives you reserve power the way they're set up.

"We're limited to thirty inches of manifold pressure, even though the engine will pull forty-four inches, but we stop it at thirty-six inches, which is basically at METO power ("Maximum, Except Take Off power" is the maximum power output to be produced by an engine—typically time limited to one to five minutes during take off)

P4Y-2G BuNo 66302 taxies in for the 2013 Planes of Fame Air Museum airshow at the Chino Airport, California. This was 66302's first appearance after having its fire bomber paint removed and the beginnings of the Coast Guard markings applied. *Nicholas A. Veronico*

PB4Y-2 Privateer Survivors

BuNo	N Number	Owner, location, notes
59701	N6884C	Hawkins and Powers, Greybull, Wyoming, ex-Tanker 127
59759	-----	Frederick Johnsen, Nampa, Idaho, cockpit section.
59819	N3739G	Lone Star Flight Museum, Galveston, Texas, ex-Tanker C30
59932	----	National World War II Museum, New Orleans, Louisana, displayed as a B-24 *Over Exposed*.
59876	6813D	Yankee Air Museum*, Ypsilanti, Michigan, ex-Tanker 125
59882	7962C	Hawkins and Powers, Greybull, Wyoming, ex-Tanker 126
66300	2872G	Yanks Air Museum, Chino, California, ex-Tanker 124
66302	2871G	4Y-2 LLC, Phoenix, Arizona, ex-Tanker 121
66304	2870G	National Naval Aviation Museum, Pensacola, Florida, incorporating parts of Buno 66261

* Note that in addition to having BuNo 59876 on display, the Yankee Air Museum holds the rear fuselage of BuNo 59905, ex-N6816D, Tanker 42, which crashed at Wenatchee, Washington. This may be incorporated into a future B-24 construction display.

for the engine on the B-25. We're never taking the engine up to full power and we still have eight inches of manifold pressure left that we're not using. We're actually being easy on the engine at the same time we have the power, so that makes it kind of nice."

It takes considerably more than money to make a four-engine warbird fly. It takes good people. "Bob West, who flew for Hawkins and Powers for years is the consummate professional," said Shoen. "He's a stickler for safety and he wrote all of H&P's safety procedures. Another guy on our team is Robert Kropp. He's a parts scrounger with no equal. He's always looking high and low for new, old stock parts, and the armament needed if we were ever to put the plane in World War II Navy configuration."

The Privateer is maintained at Dave Goss's shop Gosshawk Unlimited at Casa Grande, Arizona. "Dave Goss is one of these super-scrupulous people who doesn't throw money away, but until it is just dead right he is not happy. He is able to bring to bear the correct expertise and he is willing to contact other people and learn the ins and outs of this aircraft," said Shoen. "It's all of these people, and many others on our team, coming together that allows us to be able to fly the airplane and operate it in a safe manner. We're just stewards of history, and the Privateer is a very historic aircraft."

Many on the airshow circuit hope 4Y-2 LLC will keep the aircraft in its Coast Guard configuration and apply its livery as a tribute to America's guardians of the sea. The service's motto is *Semper Paratus*—Always Ready. And for a time in the 1950s, the Coast Guard P4Y-2G aircraft and crews were "Always Ready."

Epilogue
What Does the Future Hold?

If the past forty years have been an interesting time for the warbird community, the next forty will be even more so.

There are still dozens and dozens of World War II-era trainers that will come out of the woodwork—once-derelict Stearmans and Valiants that ended up being pushed into the backs of hangars or hung in the rafters. These warbird projects dropped off the radar screen when they were moved to home workshops, farms, or ranches, and they will one day reemerge and come onto the market.

Technology will play a greater role in putting many rare and what are today considered nonexistent types back into the air. As computer-controlled milling machines become less expensive to operate, more and more shops will have the ability to make parts from scratch. The reduction in cost of remanufacturing parts will

enable the resurrection of many aircraft that were once considered cost-prohibitive to rebuild, and will make such in-depth restorations affordable to a wider group of enthusiasts. The biggest drawback will be finding and competing for the talented craftspeople needed to operate the equipment and rebuild the aircraft.

In Japan, Nobuo Harada recovered the tail section of a Mitsubishi G4M Betty bomber (manufacturer's serial number 12017) from the South Pacific. He then attempted to recover a forward fuselage, but

One of only two original Bf-109s flying with its original Daimler-Benz DB-601 inverted V-12 engine is Werke no. 3579. This aircraft was flown by Hans-Joachim Marseille during the Battle of Britain and was known as "White 14." On September 2, 1940, Marseille belly-landed the aircraft back in France after an engagement with British Spitfires. The Luftwaffe rebuilt the aircraft as a Bf-109E-7, and it was flying on the Russian Front on August 2, 1942, when it was shot down by Hurricanes or Tomahawks. It was recovered from Russia in 1991 by Jim Pearce, and sold to the Museum of Flight in Santa Monica, California, who paid to have the German fighter restored. *Roger Cain*

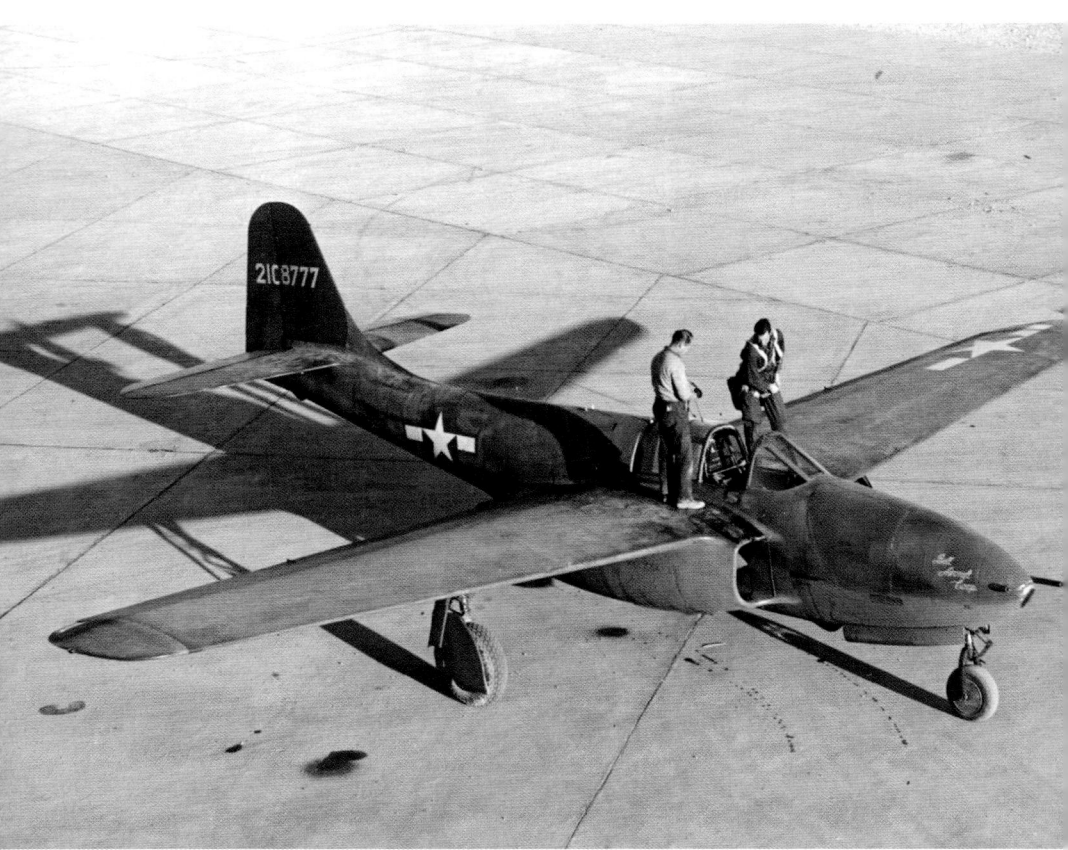

Above and Opposite: Saved from the scrap heap by Planes of Fame Air Museum founder Edward T. "Ed" Maloney in the early 1960s. When this aircraft takes wing, it will be the only flying P-59, the return of an extinct type no one thought they would ever see in the air again. Seen at the Planes of Fame Air Museum's annual open house and airshow in 2013, the YP-59 looks closer to flying than it has in a number of years. *U.S. Army Air Forces/Nicholas A. Veronico*

was unable to do so. Thus, Harada's Kawaguchiko Motor Museum, east of Tokyo, has built a forward fuselage from scratch and mated it to the original, now-restored tail section. This project is being built to static display standards, but how long will it be before the substantial remains of a G4M are brought out of the Pacific and built into a flyer? It might take funding from a deep-pocketed Japanese consortium, but it could be done, and it would be such an historic sight.

As the Zero becomes a common sight at airshows across the United States, collectors are turning their attention to other, rare

Japanese aircraft types. Today, a number of Japanese aircraft types are either under restoration to fly or are projects waiting in the wings. They include a trio of Ki-61 Tony fighters and the Planes of Fame Air Museum's Aichi D3A-2 Val dive-bomber.

While many aircraft have been rebuilt with more dependable and parts-plentiful powerplants, there is a trend where a number of restorations have gone to using rare or extinct original engine types. One of the earliest of these restorations was the Planes of Fame Air Museum's A6M-5 Zero (61-120) that has been flying with its original 1,130-horsepower, 14-cylinder Nakajima Sakae 21 radial engine. There are still a number of Sea Furies flying with their original, 18-cylinder, 3,000-horsepower, sleeve-valve, Bristol Centaurus engines, although the trend is to replace the British engine with a Pratt & Whitney R-2800. A number of Messerschmitt Bf-109s have taken to the sky with their original Daimler-Benz DB-601, 12-cylinder, inverted-V engines, and

the BMW-801 engine used to power the Flying Heritage Collection's Focke-Wolf Fw-190 is now the only one of its type in the air.

Today, the sole flying B-29 is the CAF's *FiFi*, B-29A 44-62070. It will soon be joined by Boeing-Wichita-built 44-69972, *Doc*. The first of *Doc*'s four new engines has arrived and will soon be hung. In spite of what some people believe, there is room in the skies above the United States for two or three flying B-29s on the airshow and barn-storming circuit. The addition of a second Superfortress will enable thousands more people to see an aircraft type that played such an important part in America's history.

Mustang enthusiasts can soon add another P-51B/C to the list of flying "high-backs" in the air. John Muszala's Pacific Fighters in Idaho Falls, Idaho, is working to complete an example that will be marked as Lt. William B. "Bill" Overstreet's *Berlin Express*. Among its many previous restoration projects, Pacific Fighters built-up John Sessions's P-51B *Impatient Virgin* as well as the Collings Foundation's two-seat P-51C *Betty Jane*. For years there were no early model Mustangs fly-ing, and now we have seven (*Betty Jane*, *Boise Bee*, the CAF's *Red Tail*, *Impatient Virgin*, *Ina the Macon Belle*, *Old Crow*, and *Princess Elizabeth*) and *Berlin Express* will make eight.

Although it did not see combat, Bell's P-59 Airacomet was devel-oped and built during World War II. There's one in private hands at the Planes of Fame Air Museum, and its restoration is progressing nicely. Like any other airplane, it will "fly when it flies," and when it does the aviation world will be abuzz that such a piece of history has returned to the air.

When it comes to World War II heavy iron, nothing is more iconic than the Boeing B-17 Flying Fortress. Twenty-three of the type are on public display, four are being restored for museum exhibits, six are in storage, and ten are flying today. In addition, there are five airframes under restoration to fly, albeit they are a number of years away from taking wing. That said, those five restorations speak to the health, and interest level in finding, recovering, and restoring World War II war-birds. The restoration of ex-military aircraft is a thriving industry and warbird airshows are increasing in popularity as new generations are brought to see living history in its natural environment.

The future holds a number of great aircraft recoveries and resto-rations. Some recoveries will answer questions, while others will raise more, and some types that are currently extinct will be brought to life. There are still many hidden warbirds out there waiting to be found.

Warbirds are everywhere, and here's proof of a pristine B-32 sitting in a crater, on its landing gear no less. Seeing this headline, most people scoff. On the other hand, diehard warbird enthusiasts scoff first, then pause to wonder if it could be true? Why else would humankind send people to the moon, other than to look for hidden warbirds?

Updates: One Allied, One Axis

The much-traveled B-17E *Swamp Ghost*, which was built in Seattle by Boeing, fought against the Japanese from bases in Australia, crashed in a New Guinea swamp, was recovered to California, and is now at the Pacific Aviation Museum Pearl Harbor, Oahu, Hawaii, where it will soon be displayed to the public. Swamp Ghost arrived at the museum on April 10, 2013, and plans call for the bomber to be reassembled and displayed outside in an "as-found" diorama. Once enough money is raised to restore the Flying Fortress, it will be refurbished and then moved inside Hangar 79, where it will be displayed with other aircraft telling the story of World War II in the Pacific.

The Planes of Fame Air Museum in Chino, California, continues to improve its already impressive collection of Japanese aircraft with the recent roll-out of its Judy dive-bomer. The Yokosuka D4Y *Suisei* (Comet), Allied code name "Judy," was recovered from Babo airfield, Indonesia, by Bruce Fenstermaker (see *Hidden Warbirds*, page 102), and has been restored to taxiable condition (the aircraft's longerons were repairable, but not to airworthy standards, and extruding new longerons was cost-prohibitive). In addition, the Judy has received a coat of camouflage paint. The restored dive-bomber made its public debut at the May 2013 Planes of Fame Air Show.

The iconic Pacific War combat veteran Flying Fortress known as *Swamp Ghost* arrived at its new home at the Pacific Aviation Museum Pearl Harbor, Oahu, Hawaii, on April 10, 2013. The B-17E will be reassembled in an "as-found" diorama and displayed to the public until enough money can be raised to restore the bomber and move it into Hangar 79. *Pacific Aviation Museum Pearl Harbor*

Looking much worse for wear, Yokosuka D4Y *Suisei* (Comet) was recovered from Babo airfield, Indonesia, in the early 1990s. The Planes of Fame Air Museum has restored the dive-bomber to taxiable condition, and it made its debut at the museum's May 2013 airshow. *Bruce Fenstermaker, jungle; Nicholas A. Veronico, restored*

Bibliography and Suggested Reading

Books

Alexander, Sigmund Col. *B-47 Aircraft Losses.* Spiral bound and available direct from the author (12110 Los Cerdos St., San Antonio, TX 78233-5953); newspaper clippings, reports, and photos of all B-47 losses. www.b-47.com/SAC%20Accidents.htm

Andrade, John M. *U.S. Military Aircraft Designations and Serials Since 1909.* Hinckley, Leicestershire, England. Midland Counties Publications, 1979.

Bailey, Dan E. *World War II Wrecks of the Truk Lagoon.* Redding, California. North Valley Diver Publications. 2000.

_____. *WWII Wrecks of the Kwajalein and Truk Lagoons.* Redding, California. North Valley Diver Publications. 1989.

_____. *World War II Wrecks of Palau.* Redding, California. North Valley Diver Publications. 1991.

Brandt, Trey. *Faded Contrails: Last Flights Over Arizona.* Phoenix, Arizona. Acacia Publishers, 2003.

Burtness, Robert A. *The Santa Barbara B-24 Disasters: A Chain of Tragedies Across Air, Land, and Sea.* Charleston, South Carolina. The History Press. 2012.

Cary, Alan C. *Consolidated PB4Y-2 Privateer: The Operational History of the U.S. Navy's World War II Patrol/Bomber Aircraft.* Atglen, Pensylvannia. Schiffer Military History. 2005.

Cass, William F. *The Last Flight of Liberator 41-1133: The Lives, Times, Training and the Loss of the Bomber Crew Which Crashed on Trail Peak at Philmont Scout Ranch.* West Chester, Pennsylvania. The Winds Aloft Press. 1996.

Chapman, John, and Geoff Goodall. *Warbirds Directory.* Mansfield, Nottinghamshire, England. Warbirds Media Co. Ltd. 1996. (three editions)

Childers, Thomas. *Wings of Morning: The Story of the Last American Bomber Shot Down Over Germany in World War II.* Reading, Massachussetts. Addison-Wesley Publishing Co. 1995.

Cole, Ben. *Four Down on Old Peachtree Road.* Suwanee, Georgia. Crosswind Publications, Ltd. 2007.

Coggin, Paul. *P-51 Mustang Restored: The Restoration of the Lil' Margaret. An F-6D Photo Reconnaissance Version of the P-51.* Osceola, Wisconsin. Motorbooks International. 1995.

Cupido, Joe. *Chino: Warbird Treasures Past and Present.* Riverside, California. Fox-2 Productions. 2000.

Darby, Charles. *Pacific Aircraft Wrecks . . . And Where to Find Them.* Melbourne, Australia. Kookaburra Technical Publications Party Ltd., 1979.

Dienst, John and Dan Hagedorn. *North American F-51 Mustang in Latin American Air Force Service.* Arlington, Texas. Aerofax Inc. 1985.

Doylerush, Edward. *Rocks in the Clouds: High-Ground Aircraft Crashes of South Wales.* Hersham, Surry, England. Midland Publishing/Ian Allan Publishing Ltd. 2008.

Foster, Gary Wayne. *Phantom in the River: The Flight of Linfield Two Zero One.* Ashland, Oregon. Hellgate Press. 2010.

Francillon, René J. *Japanese Aircraft of the Pacific War.* Annapolis, Maryland. Naval Institute Press. 1990.

Gallagher, James P. *Meatballs and Dead Birds: A Photo Gallery of Destroyed Japanese Aircraft in World War II*. Mechanicsburg, Pennsylvania, Stackpole Books. 2004.

Grantham, A. Kevin. *P-Screamers: The History of the Surviving Lockheed P-38 Lightnings*. Missoula, Montana. Pictorial Histories Publishing Co. Inc. 1994.

Green, Brett. *Götterdämmerung: Luftwaffe Wrecks and Relics*. London. Classic/Ian Allan Publishing. 2006.

Havener, J. K. *The Martin B-26 Marauder*. Blue Ridge Summit, Pensylvannia. Tab Books. 1988.

Hayes, David. *The Lost Squadron*. New York. Hyperion. 1994.

Hoffman, Carl. *Hunting Warbirds: The Obsessive Quest for the Lost Aircraft of World War II*. New York. Ballantine Books. 2001.

Imparato, Edward T. *Into Darkness: A Pilot's Journey Through Headhunter Territory*. Charlottesville, Virginia. Howell Press, Inc. 1995.

Job, Macarthur. *Air Disaster*. Weston Creek, Australia. Aerospace Publications Pty. Ltd. 1994. Volumes One through Four (published in 1995, 1996, 1999, and 2001 respectively).

Kearns, David A. *Where Hell Freezes Over: A Story of Amazing Bravery and Survival*. New York. Thomas Dunne Books (St. Martin's Press). 2005.

McClendon, Dennis E. *The Lady Be Good: Mystery Bomber of World War II*. Fallbrook, California. Aero Publishers. 1962.

McCurry, David L. with Cye Laramie, and Dan Thomas Nelson. *Aircraft Wrecks of the Pacific Northwest*. Bennington, Vermont. Merriam Press. 2013.

McLachlan, Ian. *Final Flights, Dramatic Wartime Incidents Revealed by Aviation Archaeology*. Haynes Publishing. London. 1995.

Macha, Gary Patric and Don Jordan. *Aircraft Wrecks in the Mountains and Deserts of California* (1909-2002, Third Edition). Lake Forest, CACalifornia. InfoNet Publishing, 2002.

Macha, Gary Patric. *Historic Aircraft Wrecks of San Bernardino County*. Charleston, South Carolina. The History Press. 2013.

Mann, Robert A. *Aircraft Record Cards of the United States Air Force (How to Read the Codes)*. Jefferson, North Carolina. McFarland & Co. Publishers, 2008.

Martin, W. W. *So I Bought an Air Force: The True Story of a Gritty Midwesterner in Somoza's Nicaragua*. Minneapolis, Minnesota. Two Harbors Press. 2013.

Martinez, Mario. *Lady's Men: The Story of World War II's Mystery Bomber and her Crew*. Annapolis, Maryland. Naval Institute Press. 1995.

Mason, Francis K. *The Hawker Typhoon and Tempest*. Bourne End, Buckinghamshire, England. Aston Publications Ltd. 1988.

Merlin, Peter W. & Moore, Tony. *X-Plane Crashes: Exploring Experimental, Rocket Plane, and Spycraft Incidents, Accidents, and Crash Sites*. North Branch, Minnesota. Specialty Press, 2008.

Mikesh, Robert C. *Broken Wings of the Samurai: The Destruction of the Japanese Airforce*. Annapolis, Maryland. Naval Institute Press. 1993.

Mireles, Anthony J. *Fatal Army Air Forces Aviation Accidents in the United States, 1941-1945* (three volumes). Jefferson, North Carolina. McFarland & Co. Publishers, 2006.

Morrison, Lee. B. *Out of the Wilderness: Restoring a Relic*. North Canton, Ohio. Military Aviation Preservation Society. 1995.

Page, Gordon. *Warbird Recovery: The Hunt for a Rare WWII Plane in Siberia, Russia*. New York. iUniverse. 2005.

Panas, Jr., John. *Aircraft Mishap Photography: Documenting the Evidence.* Ames, Iowa. Iowa State University Press. 1996.

Quinn, Chuck Marrs. *The Aluminum Trail: China, Burma, India, World War II, 1942-1945.* Self-published. 1989.

Ralph, Barry. *The Crash of Little Eva: The Ultimate World War II Survivor Story.* Gretna, Louisiana. Pelican Publishing Co. 2004.

Saunders, Andy. *Spitfire Mark I P9374: The extraordinary story of Recovery, Restoration, and Flight.* London. Grub Street Publishing. 2012.

Sheehan, Susan. *A Missing Plane.* New York. G.P. Putnam's Sons. 1986.

Shepherd, W. Richard. "They Returned: The story of the Million Dollar Valley Marauders and how they were returned to the United States for Preservation in Yesterday's Air Force." *Air Classics Quarterly Review.* Spring 1976.

Singh, Pushpindar. *Aircraft of the Indian Air Force, 1933-73.* The English Book Store. New Delhi, India. 1974.

Skaarup, Harold A. *RCAF War Prize Flights, German and Japanese Warbird Survivors.* New York. iUniverse Inc. 2006.

Smith, David J. High *Ground Wrecks and Relics: Aircraft Hulks on the Mountains of the UK and Ireland.* Leicester, England. Midland Publishing Ltd. 1997.

Spieth, Glen E. *The Swamp Ghost: B-17E 41-2446.* Unpublished manuscript. (Speith's father, Harry E. Spieth Jr. was a pilot in the 7th Bomb Group, 22nd Bomb Squadron and flew on the mission to Rabaul.)

Starks, Richard and Miriam Murcutt. *Lost in Tibet: The Untold Story of Five American Airmen, a Doomed Plane, and the Will to Survive.* Guilford, Connecticut. The Lyons Press. 2004.

Stekel, Peter. *Final Flight: The Mystery of a WWII Plane Crash and the Frozen Airmen in the High Sierra.* Berkeley, California. Wilderness Press. 2010.

Sturkey, Marion. *Mid-Air: Accident Reports and Voice Transcripts from Military and Airline Mid-Air Collisions.* Plum Branch, South Carolina. Heritage Press International. 2008.

Thompson, Scott A. *B-25 Mitchell in Civil Service.* Elk Grove, California. Aero Vintage Books. 1997.

_____. *Final Cut: The Post-War B-17 Flying Fortress and Survivors* (Third Edition). Missoula, Montana. Pictorial Histories Publishing. 2009.

Van Waarde, Jan. *US Military Aircraft Mishaps 1950-2004.* Schipol, The Netherlands. Scramble/Dutch Aviation Society. 2005.

Veronico, Nicholas A., Ed Davies, et. al. *Wreckchasing: A Guide to Finding Aircraft Crash Sites.* Castro Valley, California; Pacific Aero Press, 1992.

Veronico, N., Ed Davies, Donald B. McComb Jr., and Michael B. McComb. *Wreckchasing 2: Commercial Aircraft Crashes and Crash Sites.* Miami, Florida. World Transport Press, 1996.

Veronico, Nicholas A., Ed Davies, A. Kevin Grantham, Robert A. Kropp, Donald B. McComb, Michael B. McComb, Thomas Wm. McGarry, Enrico Massagli, and Walt Wentz. *Wreckchasing 101: A Guide to Finding Aircraft Crash Sites.* Minneapolis, Minnesota. Stance & Speed. 2011.

Veronico, Nicholas A., A. Kevin Grantham, and Scott Thompson. *Military Aircraft Boneyards.* Osceola, Wisconsin. MBI Publishing. 2000.

Veronico, Nicholas A. and Steve Ginter. *Convair PB4Y-2/P4Y-2 Privateer.* Simi Valley, California. Ginter Books. 2012.

Veronico, Nicholas A., with John M. and Donna Campbell. *F4U Corsair: Combat, Development, and Racing History of the Corsair.* Osceola, Wisconsin. Motorbooks International. 1994.

Ward, Chris and Andreas Wachtel. Dambuster *Crash Sites: 617 Dambuster Squadron Crash Sites in Holland & Germany.* Pen & Sword Books, England. 2007

Weisheit, Bowen P. *The Last Flight of Ensign C. Markland Kelly, Junior, USNR: Battle of Midway, June 4, 1942*. Baltimore, Maryland. The Ensign C. Markland Kelly, Jr., Memorial Foundation, Inc. 1993.

Wetterhahn, Ralph. *The Last Flight of Bomber 31: Harrowing Accounts of American and Japanese Pilots Who Fought in World War II's Arctic Air Campaign*. New York. Carroll & Graf. 2004.

Widner, Robert. *Aircraft Accidents in Florida: From Pearl Harbor to Hiroshima*. Lulu.com. 2009.

Wills, Richard K. *Dauntless in Peace and War: A Preliminary Archaeological and Historical Documentation of Douglas SBD-2 BuNo 2106, Midway Madness*. Washington, D.C. Naval Historical Center, 1997.

Zuckoff, Mitchell. *Frozen in Time: An Epic Story of Survival and a Modern Quest for Lost Heroes of World War II*. New York. HarperCollins Publishers. 2013.

Zuckoff, Mitchell. *Lost in Shangri-La: A True Story of Survival, Adventure, and the Most Incredible Rescue Mission of World War II*. New York. HarperCollins Publishers. 2012.

Internet Resources

Hidden Warbirds provided an extensive list of Internet resources, which has been updated and a number of sites added. Only sites with dedicated Internet addresses have been included.

Aircraft Wrecks and Aviation Archaeology

Aircraft Wrecks in the Mountains and Deserts of the American West
www.aircraftwrecks.com
Explorer G. Pat Macha's site chronicles the hundreds of sites he's documented and showcases a number of aircraft wreck mysteries he's working to solve. He is currently dedicated to locating missing WASP Gertrude Tompkins Silver, and in the process of looking for her he has located more than seventy other aircraft crash sites in the Santa Monica Bay. Macha is also the host of the History Channel's *Broken Wings* series.

Arizona Aviation Archaeology
www.aircraftarchaeology.com
Trey Brandt's excellent site on crash sites in Arizona features visits to a number of interesting wreck sites. Brandt is also the author of *Faded Contrails: Last Flights Over Arizona*.

Aviation Archaeological Investigation and Research (AAIR)
www.aviationarchaeology.com
Craig Fuller, a classically trained archaeologist, is the driving force behind Aviation Archaeology Investigation and Research, the best Internet site for ordering Army Air Force, Air Force, and Navy crash reports as well as Missing Air Crew Reports (MACRs). You can visit a number of crash sites virtually here as well.

Aviation Archaeology in Maine
www.mewreckchasers.com
Pete Noddin's Internet site is devoted to U.S. and Canadian aircraft crashes in Maine. Noddin reports that between 1919 and 1989 there were 741 military crashes in the state. His site features many of the crash sites in the state that have been investigated.

The Bent Prop Project
www.bentprop.org
Patrick Scannon, MD, Ph.D, has been traveling to the Palau Islands to search for
  downed aircraft and missing airmen from both sides of the conflict. Follow
  this outstanding work online.

Colorado Aviation Archaeology
www.coloradoaviationarchaeology.org
A sub-group of the Colorado Aviation Historical Society, this group sponsored
  the formation of the North American Institute of Aviation Archaeology
  (NAIAA), which is working to start a national organization to govern and
  develop standards for aviation archaeology.

Fatal Army Air Forces Aviation Accidents in the United States, 1941–1945
www.warbirdcrash.com
Anthony J. Mirales's site supports his comprehensive listing of more than 7,100
  stateside accidents during World War II.

Lost Flights
www.lostflights.org
Author and wreckchaser Michael B. McComb's website features many of the
  aircraft crash sites he's found in the Southwestern United States. McComb
  is best known for his explorations of crashes in the Grand Canyon area,
  and one of his galleries is devoted to aircraft accidents in this area.

Oklahoma Wreckchasing
www.okwreckchasing.org
Jeff Wilkinson's Oklahoma Wreckchasing features a great, active message board
  as well as a virtual tour of many crash sites in Arkansas and Oklahoma.

Pacific Wrecks
www.pacificwrecks.org
Justin Taylan's online database lists aircraft crashes in the battlegrounds of the
  Pacific. Taylan has also produced a number of DVDs on these crashes,
  including the Swamp Ghost and the B-17 Black Jack, as well as his
  adventures on many of the islands. Also visit Taylan's sister site: www.
  pacificghosts.com

TIGHAR—The International Group for Historic Aircraft Recovery
www.tighar.org
TIGHAR is searching for a number of high-profile aircraft crashes including
  Amelia Earhart, a TBD, and many others.

Wreckchasing
www.wreckchasing.com
This site was one of the first of the subject online in 1995 and features a very
  active message board full of tips, research information, and stories from
  people interested in the hobby. Many of today's top aviation archaeologists,
  wreckchasers, researchers, and websites can trace their enthusiasm for the
  hobby to the books *Wreckchasing: A Guide to Finding Aircraft Crash Sites*

and *Wreckchasing 2: Commerical Aircraft Crashes and Crash Sites.*

X-Hunters Aerospace Archeology Team
www.thexhunters.com
Peter W. Merlin and Tony Moore have been exploring the deserts around
Edwards AFB to honor those who perished while expanding the envelope
of flight. Their website covers every type of experimental aircraft, from
Flying Wings to Blackbirds to all of the downed X-planes. The duo has also
written a book on the subject, *X-Planes Crashes: Exploring Experimental,
Rocket Plane, and Spycraft Incidents, Accidents, and Crash Sites,* which lists
more than five hundred aircraft crash sites in the Mojave Desert area.

Warbird Restoration/Aviation Museums

Aero Trader
www.aerotrader.net
Known for their B-25 restorations, Carl Scholl and Tony Ritzman have done
everything from Mustangs and Corsairs to A-20s, A-26s, and B-26s.

The Air Museum/Planes of Fame
www.planesoffame.org
The first aviation museum west of the Mississippi River, and boasts one of the best
private collections of Axis aircraft. Annual airshow in May—not to be missed.

Airpower Unlimited
www.airpowerunlimited.net
John Lane's shop in Jerome, Idaho, has turned out a number of award winning
restorations, most recently a pair of Corsairs.

Alaska Aviation Heritage Museum
www.alaskaairmuseum.org
Located in Anchorage, the museum displays and preserves aircraft and artifacts
from the region.

American Aero Services
www.americanaeroservices.com
Just rebuilt the Collings Foundation's award-winning North American A-36A dive-
bomber, and maintains the foundation's B-17 *Nine-O-Nine* and B-24J *Witchcraft.*

American Air Museum
www.aam.iwm.uk
More than thirty thousand U.S. airmen gave their lives during World War II and
the UK remembers their sacrifice in hundreds of locations, none better
than the American Air Museum at Duxford. All of the major World War II
aircraft types are on display here.

AvSpecs Limited
www.warbirdrestoration.co.nz
Well-known for their Spitfire, Kittyhawk, Tomahawk, and Mosquito restorations.

Cal-Pacific Airmotive

www.calpacificairmotive.com

Art Teeters's shop has restored a number of the best Mustangs flying today. Cal-Pacific Airmotive holds the Limited Type Certificate for the P-51C, -51D, and -51K and provides certified parts to Mustang restorers around the globe.

Canada Aviation and Space Museum

www.aviation.technomuses.ca

Canada's National Aviation Collection holds a fantastic variety of aircraft including a number of World War II-era warbirds.

The New Canadian Air & Space Museum

casmuseum.org

Located in North York, Ontario, this museum is home to the Avro CF-105 full-scale replica.

Canadian Warplane Heritage

www.warplane.com

Certainly the largest non-government group preserving and flying World War II aircraft in Canada. Home to a fantastic museum at the Hamilton International Airport, Ontario, Canada.

Collings Foundation

www.collingsfoundation.org

Home of the Wings of Freedom air tour featuring a B-17, B-24, B-25, and two-seat P-51C. A Vietnam flight has been established with an A-4, an F-4, F-100, and Huey helicopter.

Commemorative Air Force

www.commemorativeairforce.org

One of the oldest, all-volunteer warbird flying organizations in the country, and operators of the B-29 FiFi and more than 140 other warbirds of various types and sizes.

Experimental Aircraft Association (EAA)

www.warbirds-eaa.org

www.eaa.org

Home to the EAA Warbirds and the AirVenture fly-in every July in Oshkosh, Wisconsin. They have an outstanding museum on the Oshkosh airport grounds as well.

Evergreen Aviation Museum

www.evergreenmuseum.org

Howard Hughes's HK-1 Spruce Goose greets you as you drive up to the museum, and there's a tremendous collection of warbirds as well. Check out the waterpark with the Boeing 747 slide.

Ezell Aviation

www.ezellaviation.com

Ezell Aviation has restored just about every type of warbird that has ever flown. Follow their restorations on the website.

Fagen Fighters World War II Museum
www.fagenfighterswwiimuseum.org
Home to flyable, award-winning P-38, P-40, two P-51s and a number of World
War II trainers

Fantasy of Flight
www.fantasyofflight.com
Located approximately twenty miles from Orlando, Florida, this art deco
attraction features more than forty flyable aircraft from a variety of eras.
Warbirds include a P-51C, B-26 Marauder, Short Sunderland, and many
more.

Flying Heritage Collection
www.flyingheritage.com
Warbirds from all participants of World War II painstakingly restored to as close
to original configuration as possible.

GossHawk Unlimited
www.gosshawkunlimited.com
Dave Goss's GossHawk Unlimited has rebuilt many rare aircraft, most notably
an Fw-190D, Spitfire Mk IX, Kawanishi N1K2-J George, and a Nakajima
Ki-43 Oscar. Currently working on an early variant of the Douglas A-20
Havoc, and is home to the ex-Coast Guard P4Y-2G.

Historic Flight Foundation
www.historicflight.org
Began operations in 2003 as the John T. Sessions Historic Flight Foundation and
features a number of World War II warbirds.

Imperial War Museum
www.iwm.org.uk
One of the best collections of World War II warbirds

Kalamazoo Air Zoo
www.airzoo.org
Home to an incredible aircraft collection and restoration center that has recently
restored an SBD and the Curtiss XP-55 Ascender.

Legend of Aces
www.legendofaces.com
Warbird restoration and maintenance, and home of the *Sandbar B-25 Mitchell*.

Lewis Air Legends
www.lewisairlegends.com
More than two dozen aircraft ranging from the P-38F *Glacier Girl* to the P-39
Brooklyn Bum, a pair of F7F Tigercats, as well as four F8F Bearcats.

Liberty Foundation
www.libertyfoundation.org
Home of the Salute to Veterans B-17 tour with another B-17G under rebuild.

Mid-Atlantic Air Museum P-61
www.maam.org
Follow the P-61 Black Widow restoration at the museum's website.

Midwest Aero
www.midwestaero.com
Numerous award-winning Mustang restorations have come from Midwest Aero
including *Lil' Margaret*, *Happy Jack's Go Buggy*, *Daddy's Girl*, and *Was That
Too Fast*.

Military Aviation Museum
www.militaryaviationmuseum.org
Home to one of the largest private collections of warbirds from all sides of the
war, everything from a Polikarpov I-16 and Hawker Hurricane to a Curtiss
P-40 and a B-17G Flying Fortress.

Museum of Flight
www.museumofflight.org
The museum acquired the Champlin Collection and based its new Personal
Courage Wing around its collection of World War II and II aircraft.

National Air & Space Museum
www.airandspace.si.edu
This nation's collection of history-making aircraft with locations in downtown
Washington, D.C., and the Steven F. Udvar-Hazy Center in Chantilly,
Virginia.

National Naval Aviation Museum
www.navalaviationmuseum.org
The U.S. Navy's repository of historic and record setting aircraft.

National Museum of the U.S. Air Force
www.nationalmuseum.af.mil
Air Force aircraft from the World Wars to supersonic test birds displayed at
Wright-Patterson Air Force Base, Ohio.

Pacific Aviation Museum
www.pacificaviationmuseum.org
Warbirds from World War II, Korea, and the Vietnam War on display at Ford
Island, Oahu, Hawaii. World War II focus in on the Pacific Theater.

Pacific Fighters
www.pacificfighters.net
Warbird restoration and maintenance, known for their P-51B and C model
restorations including *Betty Jane* and *Impatient Virgin*.

Pima Air & Space Museum
www.pimaair.org
More than three hundred aircraft on display with plenty of World War II
warbirds. This is also the place for tours of the U.S. military's boneyard,
known as AMARG, at Davis-Monthan AFB.

Pioneer Aero
www.pioneeraero.co.nz
Restorations of an Lavochkin La-9, Yak-3, Corsair, and now working on a
Hawker Tempest II.

Royal Air Force Museum (London and Cosford)
www.rafmuseum.org.uk
Britain's only national museum dedicated entirely to aviation.

Sanders Aeronautics
www.sandersaircraft.com
Warbird restorations on everything from Sea Furies, to Corsairs, to an Me-262
replica.

San Diego Aerospace Museum
www.sandiegoairandspace.org
Warbirds in the collection include a Spitfire Mk XVI and Grumman F6F
Hellcat.

Stallion 51
www.stallion51.com
P-51 flight training in dual control Mustangs.

The Fighter Collection
www.fighter-collection.com
Europe's largest collection of flyable World War II aircraft.

Tillamook Air Museum
www.tillamookair.com
More than thirty warbirds on display, nearly all flyable including a P-38, P-51,
F4U-7, J2F-6 Duck, Bf-109, AM-1 Mauler, and many more. The museum
will move to Madras, Oregon, in late 2015, early 2016, and the name will
change to The Erickson Aircraft Collection.

Vintage Aircraft
www.twinbeech.com
Restorers of the Twin Beech series and providers of the Aerospace Products spar
strap kit. Restorers and operators of an eight-gun nose PV-2D Navy patrol
bomber.

Vintage V-12 and Vintage Radials
www.vintagev12s.com
www.vintageradials.com
Two shops: FAA certified repair stations operated by Mike Nixon and known for
Allison, Merlin, and Griffon rebuilds, as well as the BMW801 radial now
flying in the Flying Heritage Collection's Fw-190.

Vulture's Row Aviation
www.vulturesrowaviation.com
U.S. Navy tailhook equipped warbirds are their specialty. Currently restoring an
SBD and SB2C side-by-side.

Warhawk Air Museum

www.warhawkairmuseum.org

Home to the P-51B *Boise Bee*, P-51D *Hell-Er-Bust*, and P-40E *Sneak Attack*, and P-40N *Parrot Head* among other warbirds.

Westpac Restorations

www.westpacrestorations.com

Currently working on a number of P-38s, P-47s, an F7F and an F4U. Co-located with the National Museum of World War II Aviation.

Yanks Air Museum

www.yanksair.com

Home to a large collection of World War II warbirds from the P-51A and C-47 to an F6F-5 and PB4Y-2.

1941 Historical Aircraft Group Museum

www.1941hag.org

The museum is currently restoring a Douglas B-23 and A-20 Havoc.

Warbirds, Air Racing

A. & T. Recovery

www.atrecovery.com

A. & T. Recovery specializes in the underwater recovery of historically significant aircraft. See what they've brought up, and learn about what they're going after.

AAFO—All Aviation Flightline Online

www.aafo.com

Air racing, warbirds, airshows, photo galleries, message board, flight simulators, all in one place.

B-17 Flying Fortress and B-25 Mitchell

www.aerovintage.com

Author Scott Thompson's site for news and information on both the B-17 and the B-25. There's some interesting pages on the Tallmantz operation, B-17s in the movies, aircraft disposed of through the War Assets Administration, and pages for the restoration of the B-17E *Desert Rat/Tangerine* (XC-108A).

B-17E *Swamp Ghost*

www.theswampghost.com

www.aeroarchaeology.com

Two sites dedicated to the preservation and recovery of the B-17E *Swamp Ghost*. The Swamp Ghost is operated by Justin Taylan, who had made dozens of trips to the bomber and produced a video about its history and crash. Aero Archaeology is Fred Hagen's website detailing his recovery of the B-17 and other aircraft.

B-17G *Lacey Lady*

www.b17alliancegroup.com

Art Lacey flew a B-17 home from the scrapyard and put it on display above his gas station and motel complex in Milwaukie, Oregon. Follow along as the family works to restore this Flying Fortress.

B-24 *Lady Be Good*

www.376hbgva.com/aircraft/ladybegood.html
www.ladybegood.com
Visit the bomb group's site or ladybegood.com for author Mario Martinez's site on the aircraft and the myths surrounding it loss and discovery.

B-24 Liberator Australia

b24australia.org.au
Restoring the only surviving intact B-24 in Australia.

Courtesy Aircraft

www.courtesyaircraft.com
Mark Clark has been in the warbird sales business for more than thirty-five years and has sold more than 2,500 warbirds, many multiple times.

Hawker Tempest

www.hawkertempest.se
Christer Landberg's site for all things Tempest.

National Aviation Heritage Invitational

www.heritagetrophy.com
Home website for information and coverage about the Neil A. Armstrong Aviation Heritage Trophy and its yearly competitions held every September in conjunction with the National Championship Air Races in Reno, Nevada.

P-51 Mustangs

www.mustangsmustangs.com
Mustang automobiles or P-51 Mustang airplanes, both are found at this website. Up-to-the-minute news on P-51 restorations, history, ace interviews, and much more.

Swiss Mustangs

www.swissmustangs.ch
Martin Kyburz's site started out focusing on the Swiss Air Force use of the P-51 and what happened to the planes when they were phased out of service. The site has expanded to cover nearly every aspect of the Mustang's service.

Mustang!

www.mustang.gaetanmarie.com
Gaetan Marie's website dedicated to the P-51 with galleries on many unique uses of the P-51.

XP-51G *Margie Hart*

www.xp51g.com
John Morgan's site to follow the restoration of the sole surviving XP-51G.

P-38 *Glacier Girl*

p38assn.org/glacier-girl.htm
The P-38 national association's webpage dedicated to *Glacier Girl*.

Platinum Fighter Sales

www.platinumfighters.com
Own a piece of history! Platinum Fighter Sales has handled some of the highest profile warbird transactions in recent history including P-38F *Glacier Girl*.

National Championship Air Races

www.airrace.org

Information about upcoming races in Reno, Nevada, plus performer profiles, racing statistics, photos, and videos of air racing past and present. Six classes of air racers turn the pylons every September, some going more than 500 mph. Bucket list item if you've never been.

Society of Air Racing Historians

www.airrace.com

Everything you ever wanted to know about the history of air racing. The group holds an annual symposium each year that attracts many pilots and crewmembers.

Society for Aviation History

www.sfahistory.org

Aviation history presented through educational programs and newsletters. Five programs per year, typically presented by those who made aviation history.

Warbird Aero Press

www.warbirdaeropress.com

Hosted by Scott Germain, this site covers "the art, action, and adrenaline of Unlimted Air Racing." Stories, interviews, art, and photography of all aspects of the Unlimiteds.

Warbirds Directory

www.goodall.com.au

Geoff Goodall's Warbirds Directory has gone electronic because it is just too large to sell as a hard-copy version. Goodall has generously put Volume Six of the directory on his site for free.

Warbird Information Exchange or WIX

warbirdinformationexchange.org/phpBB3/index.php

Message board for all things warbirds and includes the Warbird Registry giving the history of most surviving ex-military types.

Worldwide Aircraft Recovery, Ltd.

www.worldwideaircraft.com

You name it, they've moved it, from P-51s to the SR-71.

XP-82 Project

www.xp-82twinmustangproject.com

Restoration project led by Tom Reilly. Website has monthly updates on the aircraft's rebuild to flight program.

Aviation Magazines with Warbird Coverage

AAHS Journal (American Aviation Historical Society member magazine)
www.aahs-online.org

Aeroplane magazine
www.aeroplanemonthly.com

Air Classics / Warbirds International / Mustangs
www.challengeweb.com

Air & Space/Smithsonian magazine
www.airspacemag.com

Aviation History
www.historynet.com/magazines/aviation_history

CAF Dispatch (Commemorative Air Force member magazine)
www.commemorativeairforce.org

Classic Wings
www.classicwings.com

Flight Journal
www.flightjournal.com

Flight Path
www.yaffa.com.au/index.php/consumer-publications/flightpath

FlyPast
www.flypast.com

In Flight USA
www.inflightusa.com

Touch & Go (Society for Aviation History member newsletter)
www.sfahistory.org

Warbird Digest
www.warbirddigest.com

Warbirds Magazine (EAA Warbirds member magazine)
www.warbirds-eaa.org/magazine

Warbirds News (on-line warbird magazine)
www.warbirdsnews.com

Index